Watching Weather

Tom Murphree and Mary K. Miller with the Exploratorium

An Exploratorium Book

An Owl Book
Henry Holt and Company   New York

Henry Holt and Company, Inc.
Publishers since 1866
115 West 18th Street
New York, New York 10011

Henry Holt® is a registered trademark of
Henry Holt and Company, Inc.

Published in Canada by Fitzhenry & Whiteside Ltd.,
195 Allstate Parkway, Markham, Ontario L3R 4T8.

Library of Congress Cataloging-in-Publication Data
Murphree, Tom.
    Watching Weather/Tom Murphree and Mary K. Miller with
the Exploratorium.
        p. cm. — (Accidental scientist) (Exploratorium book)
    "An Owl book."
    ISBN 0-8050-4542-2
    1. Weather—Popular works. 2. Weather—Folklore. I. Miller, Mary K.
II. Exploratorium (Organization) III. Title. IV. Series.
V. Series: Exploratorium book.
QC981.2.M55 1998   97-49801
551.5—dc21   CIP

Henry Holt Books are available for special promotions and
premiums. For details, contact: Director, Special Markets.

First American Edition 1998

Designed by Gary Crounse

Be careful! The experiments in this publication were designed with
safety and success in mind. But even the simplest activity or most
common materials can be harmful when mishandled or misused.

The Exploratorium® is a registered trademark and service mark of The
Exploratorium.

ChapStick® is a registered trademark of A.H. Robins Company, Rich-
mond, Virginia.
Mickey Mouse© is a copyright of The Walt Disney Company.

Printed in the United States of America
All first editions are printed on acid-free paper. ∞

10 9   8   7   6   5   4   3   2   1

Contents

Other Titles in The Accidental Scientist Series

The Garden Explored by Mia Amato with the Exploratorium

The Inquisitive Cook by Anne Gardiner and Sue Wilson with the Exploratorium

The Sporting Life by Susan Davis, Sally Stephens, and the Exploratorium

Children's Books from the Exploratorium

The Science Explorer: Family Science Experiments from the World's Favorite Hands-On Museum by Pat Murphy, Ellen Klages, Linda Shore, and the Exploratorium

The Science Explorer Out and About: Fantastic Science Experiments Your Family Can Do Anywhere by Pat Murphy, Ellen Klages, Linda Shore, and the Exploratorium

Other Books for Adults from the Exploratorium

The Color of Nature by Pat Murphy and Paul Doherty

By Nature's Design by Pat Murphy

Introduction

Welcome to The Accidental Scientist, a series of books created by the Exploratorium to help you discover the science that's part of things you experience every day.

In *Watching Weather*, we investigate the science of summer days and winter winds, of heat waves and cold snaps, of changes in the weather and ways you can see them coming. We tell you about the basic science that explains the vagaries of the weather. Why are desert nights so cold? Why is it hotter downtown than it is in the nearby countryside? What do the clouds tell you about coming changes in the weather? We give you rules of thumb that will help you predict your local weather—and we explain the science behind these rules.

Most weather books focus on your local weather. But the weather outside your front door is connected to the weather everywhere else in the world. Your local weather is linked to and driven by conditions thousands of miles away. To understand the weather in your neighborhood, you need the global perspective this book offers.

Be warned: At the Exploratorium, San Francisco's museum of science, art, and human perception, we've found that once you start watching the weather, you may find it difficult to stop. While we were working on this book, the Exploratorium staff became obsessed with observing indications of changing weather. The copy editor would return from lunch and warn people that a cold front was coming in. How did she know? From examining the clouds.

After reading this book, you may find yourself aware of things that you never paid any attention to before, and asking questions that you never thought to ask. Have fun!

Goéry Delacôte
Director
Exploratorium

1 *Earth & Sun*

The big picture

THE METEOROLOGICAL HALF OF OUR WRITING TEAM, Tom Murphree, defines weather as the state of the earth's atmosphere. A glance out your window gives you a local view of the current state of the atmosphere—it's windy or calm; it's cloudy or clear. The evening news offers a global view of the atmosphere, using images and information provided by weather satellites. Meteorologists use both the local view and the global view to predict how your local weather is likely to change over the next few days.

To make sense of weather, you need both of these perspectives—the global view and the local view. The snowstorms that shut down airports on the East Coast, the heat waves that send temperatures soaring in the Midwest, and the cold snaps that ruin crops in Florida and California—all of these local changes in the atmosphere result from global disturbances, which have resulted, in turn, from local changes at distant locations.

Conveniently, the scientific principles that explain local changes also apply to global disturbances. When you understand what causes your local breezes to blow, you'll have a handle on what creates the powerful, high-altitude winds known as *jet streams*. In this book, we'll explain the basic scientific concepts behind weather and use these concepts to describe some rules of thumb that will help you predict your local weather.

A small fluctuation in local temperature can lead to a major change in the weather halfway around the planet a week later.

Differences in temperature between one place and another are an underlying cause of the shifts in the atmosphere that we experience as changes in the weather. A small fluctuation in local temperature can lead to a major change in the weather halfway around the planet a week later.

To understand your local weather, you need to know how the earth is warmed and cooled and why some spots are warmer than others. Viewed from a global perspective, the information can help you understand how the weather patterns where you live are linked to weather patterns around the world.

Tropical Heat and Arctic Chill

Light from the sun provides most of the energy that warms our planet. Solar energy drives the weather—the sun's energy creates winds, evaporates water that later falls as rain, and provides the power that propels hurricanes and tornadoes. Taking a look at the big picture—the relationship of the earth to the sun—is crucial to understanding what makes one location cold and cloudy while another roasts in intense sunlight.

Tom Murphree lives in Monterey, California, a foggy town on the cold eastern edge of the north Pacific Ocean. On sunny days, he tries to take a break from his work and get outside. For his cat, Zuchar, making time for sunshine is not a problem. If the sun shines during the morning, Zuchar is sure to be spread out on the east-facing part of the roof of Tom's house. If the sun comes out around lunchtime, Zuchar will be lounging on the south side of the roof. And if the sun breaks through the fog late in the afternoon, he'll be on the west-facing roof.

Whether they know it or not, heat-seeking tourists who head for tropical regions are actually trying to get the best possible angle on the sun.

Tom's cat is a solar engineer who's figured out all the angles. When the sun comes out, he heads straight for the side of the roof that faces the sun most directly. He doesn't bother with the flat roof of the garage, which can't tilt him toward the sun the way the sloping house roof does. He also knows how to maximize his exposure by laying himself out broadside to the sun's rays. Zuchar intuitively knows that to get the most from solar energy, you've got to turn toward the sun and let its rays hit you straight on and all over. Zuchar's solar tracking reveals a lot about what makes some places and times of the year hotter than others.

Whether you are a heat-seeking cat or a snowbird tourist in search of a warm vacation spot, a good strategy is to look for the place that gets the most direct sunlight. When the sun is shining straight down, you get a concentrated dose of solar energy. If the sun is low in the sky, that same energy hits at a more glancing angle and is spread over a larger area, providing less heat per square inch.

From the point of view of a heat-seeking cat, looking for direct sunlight means looking for a roof that's tilted toward the sun. Others make use of the same principle. Dedicated sunbathers tilt their lawn chairs so that the sun shines on them more directly. Farmers sometimes plant crops that require more sunlight on hills that slope toward the sun. Energy-conscious architects orient buildings so that sunlight will provide warmth in the winter.

If you're operating on a global scale, the search for direct sunlight means looking for a place where the earth's surface is tilted toward the sun, which means choosing the right latitude. Tourists from temperate regions of North America often head to the tropics to escape winter. Whether they know it or not, those heat-seeking tourists are trying, like Zuchar, to get the best angle on the sun. Tropical regions receive very direct solar radiation. To the north and south of the tropics, the earth's surface tilts farther away from the sun's rays and sunlight comes in at a lower angle.

The tropics lie between the Tropic of Cancer, 23½° north of the equator, and the Tropic of Capricorn, 23½° south of the equator. If you are right on the Tropic of Cancer or the Tropic of Capricorn, the sun is directly overhead once a year. If you live inside the region

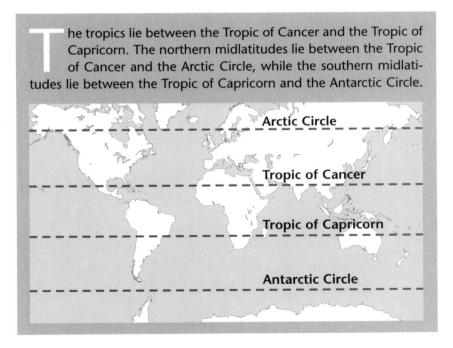

The tropics lie between the Tropic of Cancer and the Tropic of Capricorn. The northern midlatitudes lie between the Tropic of Cancer and the Arctic Circle, while the southern midlatitudes lie between the Tropic of Capricorn and the Antarctic Circle.

Arctic Circle

Tropic of Cancer

Tropic of Capricorn

Antarctic Circle

Reasons for the Seasons

The earth's axis, the imaginary line through the earth around which our planet spins, is tilted with respect to the earth's orbit around the sun.

In the Northern Hemisphere, the shortest day of the year is on about December 21, the day of the winter solstice. On this day, the earth is at a point in its orbit where the Northern Hemisphere is most tilted away from the sun, and the Southern Hemisphere is most tilted toward the sun. If you are anywhere north of the equa-

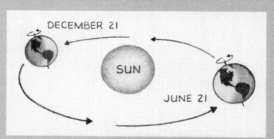

tor, the noon sun will be at its lowest point in the sky on that day. If you're standing on the Tropic of Capricorn, the noon sun will be directly overhead. When the earth is in this posi-
tion, sunlight shining on a particular location in the Northern Hemisphere is less direct than it is at any other time of year, and the days are shorter. So the Northern Hemisphere is cooler and experiences winter. In the Southern Hemisphere, sunlight is more direct and the days are longer. So the Southern Hemisphere is warmer and has its summer.

Six months later, on June 21, the earth is at the point in its orbit where the Northern Hemisphere tilts most toward the sun and the Southern Hemisphere tilts most away. This is the day of the northern summer solstice, the longest day of the year in the Northern Hemisphere. If you're standing on the Tropic of Cancer, the noon sun will be directly overhead. If you are anywhere to the north of the equator, the noon sun will be at its highest point in the sky. This is the beginning of the Northern Hemisphere's summer and the Southern Hemisphere's winter.

defined by these boundaries, the sun is directly overhead twice a year. Outside this region, the sun is never directly overhead.

What's so magical about 23½°? Well, maybe you've noticed that most globes are mounted to spin around a tilted axis. That's to show that the earth's axis, the imaginary line through the earth around which the planet spins, is not perpendicular to the plane in which the earth travels as it orbits the sun. Instead, the earth is tilted at an angle of 23½° relative to its orbit. Throughout the year, the earth's North Pole, the northern end of its axis, points toward Polaris, the North Star. This is why the North Star appears to remain stationary in the night sky of the Northern Hemisphere.

The earth's tilt affects how much sunlight shines on different regions of the earth at different times of the year. That, in turn, creates the seasons. Because of the earth's tilt, the Northern Hemisphere leans toward the sun in June and away from the sun in December. When the sun is directly overhead, the sun's rays hit the ground at a 90° angle. In June, the sun's rays shine on the Northern Hemisphere more directly, at an angle closer to 90°, than in December. In addition, the days are longer. The Northern Hemisphere absorbs more solar energy and warms up.

In December, the sun's rays strike the Northern Hemisphere at a more glancing angle, spreading the solar energy over a larger area. The days are shorter. The Northern Hemisphere absorbs less solar energy.

Your position on the planet dictates both how directly and how long the sun shines down on you each day. On June 21, if you're on the Tropic of Cancer, the sun is directly overhead, 90° above the horizon, and the day will be about 13½ hours long. Suppose you're just a little bit farther north, at San Diego, California, or Jacksonville, Florida, which are both located at about 30° north latitude. The noon sun will be a little lower in the sky, 83½° above the horizon. But that latitude stays in daylight a bit longer, getting about 14 hours of sun. Move to 60° north latitude, in Anchorage, Alaska, and the noon sun will be even lower in the sky, just 53½° above the horizon. But the day lasts 18½ hours. Barrow, Alaska, at about 70° north latitude, has daylight that lasts for two continuous months, but the noon sun is just 43½° above the horizon.

The Temperature Story

On an annual average, places closer to the equator, where the sun gets highest in the sky, tend to be warmer than places farther to the north or the south. That explains why Key West, Florida, has an annual average temperature of 78°F, while International Falls, Minnesota, has an average annual temperature of just 36°F. International Falls may have longer days in the summer, but it has shorter days in the winter and less direct sunlight throughout the year. So its average temperature is much lower than Key West's.

But what's true for annual averages can be reversed when you look at seasonal temperatures. Compare, for example, San Diego, California, and Sioux Falls, South Dakota. San Diego, in southern California, is a popular tourist town, famous for palm trees, sandy beaches, and warm weather. It's much closer to the equator than Sioux Falls, a city on the northern Great Plains, located in a land of broad prairies and bitterly cold winters. In midwinter, San Diego is definitely warmer than Sioux Falls. But in midsummer, the average daily high in Sioux Falls is in the mid-80s, while the average in San Diego is more than 10 degrees colder.

Yatkusk, a Siberian city, has one of the greatest ranges of annual temperatures in the world, a range of 62.2°C. It's at 60°N and far from water.

To understand variations in temperature, you need to consider both the angle of the sun and the length of the day. Longer days can compensate for a sun that's lower in the sky. In the summer, the days are longer in Sioux Falls than in San Diego, so Sioux Falls gets warmer—even though the sun is lower there.

But there's still more to the story. You've seen that how much solar energy the earth receives varies from place to place and from season to season. How that solar energy affects local temperature also depends on local conditions in the atmosphere and on the earth's surface. San Diego is a coastal city, a factor that dramatically affects its temperature swings.

In the Northern Hemisphere, June is the month with the highest number of daylight hours. So why is August hotter?

The time lag between when the Northern Hemisphere gets the most sunlight and when it's the warmest results from the earth's capacity to absorb and hang on to solar energy. The water, land, and air can store this energy before releasing it back into space. As long as more energy is coming in from the sun than is lost by the earth, the earth warms up. It's like your bank account: If you keep depositing more money into the account than you withdraw, your balance increases. Starting from the coldest days of winter (around January) until the hottest days of summer (around late July), more energy comes into the Northern Hemisphere than leaves, so temperatures rise. The balance shifts around the middle of August and temperatures begin to fall because the Northern Hemisphere begins losing more energy than it receives.

The reverse is true for the winter solstice, around December 21. The temperatures typically continue to drop in the Northern Hemisphere from the winter solstice until February, even though daylight hours are increasing. During this time, the Northern Hemisphere loses more energy than it gains from the sun.

If the earth did not hang on to any of the sun's energy and release it over time, the average temperature in the Northern Hemisphere might look like the white line on the graph below—hottest on June 21 and coldest on December 21. The black line shows the actual average temperature.

J F M A M J J A S O N D

A Walk Along the Beach

If you've ever walked barefoot on the beach on a summer day, you've probably noticed that at midday, when the sand is hot, the water is blessedly cool. Walk the same beach at night, and you'll find that the sand has cooled off, but the water temperature is about the same as it was at midday, and is now warmer than the sand.

Land warms up faster and can reach higher temperatures than water. It also cools off faster and can reach lower temperatures. As a result, variations in air temperature are much greater over land than over water. If you are looking for temperature extremes, go inland. If, on the other hand, you are looking for a more moderate climate, head for the water. On a sunny summer day, you'll often find a cool breeze by a lake, a river, or an ocean. (See page 29.) On a local scale and on a global scale, water provides a moderating influence on air temperature.

Water tends to warm up more slowly than land. Rather than just warming the surface of the water, sunlight shines through the surface and warms deeper waters as well. Wind blowing over the water may mix warm surface water with cooler, deeper water. Currents may carry warm water away and bring in cooler water. As the water mixes and flows, its warmth is spread over a large volume. Over the course of a day, energy absorbed from sunlight can change the temperature of a body of water to a depth of 20 feet or more.

On land, however, the solar energy absorbed during a day might only warm the ground to a depth of about a foot. The same amount of energy may be absorbed by the water and the land. But on the land, the energy is concentrated near the surface, so the land's surface temperature tends to be higher. Dig your feet down into the hot sand next time you're at the beach and you'll find cooler sand not far down.

Water has a remarkable ability to absorb and hold a lot of energy before it gets hot. If you've ever tried to eat a pizza fresh from the oven, you've probably noticed that you can nibble on the bare crust without burning yourself. But if you try to take a bite loaded with tomato sauce, you're likely to regret it. Both the crust and the sauce are at the same temperature, but the sauce, with its

Heat and Temperature— What's the Difference?

People often use the words "heat" and "temperature" interchangeably, but they're really two different things.

Heat is a form of energy that flows from a hotter object to a cooler one. Heat flowing out of an object causes that object's temperature to drop, while heat flowing into an object causes that object's temperature to rise.

But here's the tricky part: If the same amount of heat flows into two different things, their temperatures won't necessarily increase by the same amount. The amount the temperatures change will depend on several factors.

A major factor is the material. Different materials respond differently to heat. For example, just a little heat raises the temperature of most metals very quickly. On the other hand, it takes a much larger amount of heat to raise the temperature of water.

Another big factor is how much of the material you are heating. The more there is of something, the more heat is required to raise its temperature. Imagine pouring a cup of boiling water into a saucepan of cold water. Now imagine pouring a cup of boiling water into a bathtub of cold water. Though you're adding the same amount of heat to the cold water in the saucepan and in the bathtub, their temperatures change by different amounts. The temperature of the water in the saucepan rises noticeably from cold to lukewarm, but the bath water still feels cold.

The ocean is like an enormous bathtub. Huge amounts of heat are needed to change its temperature even slightly. That's one reason the temperature of the ocean remains relatively constant throughout the year.

high water content, absorbed several times more energy from the hot air in the oven. When this extra dose of energy is transferred from the sauce to your tongue, you get burned.

Water has a related ability to lose a lot of heat while its temperature goes down a relatively small amount. That's why the pizza sauce stays warm long after the crust has cooled to room temperature. Scientists describe these two abilities by saying that water has a high *heat capacity*. That is, water has a high capacity to absorb or release heat while undergoing some standard change in temperature. If you want to increase the temperature of a pound of water and a pound of rock by one degree Fahrenheit, it will take about three times as much heat to increase the temperature of the water as it will to increase the temperature of the rock.

Because of its high heat capacity, water warms up more slowly than land and also cools more slowly. The land stores its warmth near the surface where it can easily be released. Water, on the other hand, stores much of its warmth below the surface. As the surface water cools, it sinks and is replaced by water from below. The rising water is warmer than the water it replaced, so the surface cools down only slightly. In addition, the higher heat capacity of water means that the same loss of energy from water and land causes a small temperature drop at the water's surface but a large drop at the land's surface. So the land cools quickly while the water cools slowly.

The moderating influence of water helps explain why the weather of many coastal cities (such as San Diego) tends to be milder, lacking the great temperature swings of land-locked areas (like Sioux Falls). This is particularly true of coastal cities where winds usually blow from the ocean to the land, cooling off the land.

Ocean Motions

Most tourists coming to California from the East Coast in the summer expect much warmer ocean waters than they find. Hollywood is partly responsible for that; movies about California surfers and beach parties have fueled the misconception that we Californians spend all our time frolicking in the waves. But folks from the East Coast also expect a warm summer ocean because the Gulf Stream, an ocean current that flows northward along much of the eastern coast of North America, brings warm waters to their shores. Here in California, we both suffer and benefit from the California Current, which flows from the chilly North Pacific down along the Oregon and California coasts. This band of cold surface water helps keep typical water temperatures along the shore in the chilly 55–65°F range, making ocean swimming nearly unbearable unless you wear a rubber wetsuit.

Ocean currents don't just affect the comfort of beachgoers. The warming and cooling effects of warm and cold ocean currents influence weather throughout the world. Currents flowing toward the tropics from cooler regions of the ocean transport cool water, which cools the overlying air. Currents flowing away from the tropics move warm water into cool regions and warm the overlying air.

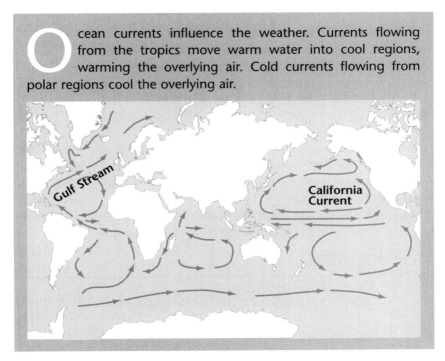

Ocean currents influence the weather. Currents flowing from the tropics move warm water into cool regions, warming the overlying air. Cold currents flowing from polar regions cool the overlying air.

Urban Hot Spots

During the day, air temperatures in some cities can be around 5°F warmer than nearby rural areas; the difference can be even greater at night. In infrared satellite images that show the temperature of regions of the earth, cities stand out as hot spots, or urban heat islands.

Several factors combine to make a heat island. People and their cars, trucks, and other machines generate heat. Naturally, that heat is most concentrated where people congregate.

The materials used to build cities also contribute to the creation of a heat island. Asphalt streets reflect only about 5 percent of the sunlight shining on them; the rest goes to warm the asphalt. A grassy meadow, by comparison, reflects three to five times more sunlight than asphalt, and stays cooler. No wonder asphalt can be 36°F hotter than nearby grass, hot enough to fry an egg.

On a summer day, not just the streets and sidewalks but the sides of buildings, too, can be hot enough to fry an egg. When you walk down the street, infrared radiation from the pavement and the buildings surrounding you can make you feel like you're being cooked.

The lack of water and plants in a city adds to urban warmth. When water evaporates from the leaves of plants or the surface of a lake, it cools the surrounding air. In some cities, planners are trying to offset urban warming by creating more city parks, especially parks with ponds and lakes.

To determine whether your city is a heat island, take some temperature measurements, starting in the countryside and moving in toward downtown. Make four or five stops along the way and measure the temperature in each spot. Make sure that you make each measurement in the shade and at the same height above the ground as previous measurements. Then turn around and go back again, stopping in the same places. The best time to try this is on a summer evening, because that's when the temperature differences are likely to be greatest.

Compare, for instance, summer temperatures in Los Angeles, California, and Charleston, South Carolina, cities that are at about the same latitude. In July, Los Angeles has an average high air temperature of 75°F, while Charleston's is 90°F. This difference is due, in part, to the difference in temperature of each city's nearby ocean. Off Los Angeles, the California Current helps to keep the ocean waters relatively cool, in the mid-60s. Off the coast of South Carolina, the northward-flowing Gulf Stream allows the ocean waters to stay in the upper 70s.

Weather from the Ground Up

Walking barefoot on the beach can give you a feel for ocean temperatures. Walking barefoot away from the beach and back to your car can teach you—the hard way—that not all surfaces on land warm up at the same rate. Some surfaces get so hot they'll burn your feet, while others are pleasantly warm, and still others are actually cool.

For example, you've probably learned that if you're going to cross a parking lot in bare feet on a hot summer day, you'd be wise to walk on the painted white lines. Dark surfaces, like black asphalt, reflect only a little of the light shining on them. Most of the light is absorbed and warms the surface. Light-colored surfaces, like the painted white lines, reflect much of the light they receive—that's what makes them look light-colored. Since they absorb less of the light, they don't warm up as much. That's one reason that light-colored clothing is popular in the summer. Light colors reflect more sunlight than dark ones, which helps light-colored clothing stay cooler.

If you're walking barefoot over bare dirt or sand, you'll probably find yourself scampering to get to the next patch of grassy ground. Grass and the soil underneath it can hold a lot of water. Water, as we mentioned before, warms up more slowly than rock or soil. If a square foot of dry ground and a square foot of wet ground absorb the same amount of sunlight, the wet ground ends up cooler.

Not only does water warm up slowly, it also evaporates, with warm water evaporating more than cool water. Water can evapo-

Earth's Energy Budget

Sunlight helps warm the earth, and infrared radiation lost to space helps cool the planet down. Together, these inputs and outputs of energy help determine earth's overall temperature.

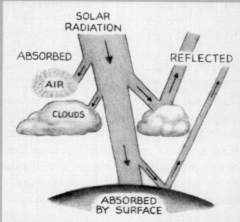

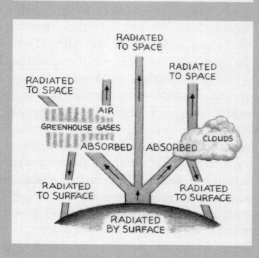

Incoming solar radiation

Sunlight shines on the earth; clouds, air, and the planet's surface reflect some of this energy and absorb some of it.

Outgoing radiation

The earth's surface releases the absorbed energy. Much of that released energy is in the form of infrared radiation. Some of this energy is absorbed by clouds and the air, particularly by the greenhouse gases. When the clouds and gases release this energy, some of it radiates out to space—but some returns to the surface, helping to keep the earth warm.

rate out of the pores of the grass leaves and out of the cracks and little holes that extend down into the soil. Evaporation cools things down, which is why you cool down when your sweat evaporates. So although dark green, grassy ground may absorb a lot of sunlight, it has an evaporative cooling system that can keep it cooler than lighter-colored, but dry, ground.

The kind of temperature differences that you might experience when you are barefoot can help create *microclimates,* small areas with climates that differ from their surroundings. On a hot day in the city, for example, you may find a region of cool air in a park, where green plants are helping to keep things cool. On a cold winter day, you may find a warm spot by a dark brick wall that's been warmed by the sun and is radiating the energy it has absorbed. The way that the greenery and the brick deal with the energy of sunlight affects the temperature of their immediate surroundings.

The temperature differences created by different surfaces can affect weather on a larger scale as well. The creation of cities, with their many paved surfaces, has altered the weather in certain areas, creating urban *heat islands* (see page 14) where the weather is warmer than it is in the surrounding countryside.

A Blanket of Air

Conditions on the ground affect the temperature of the air just above the ground—and so do conditions in the air itself. Have you ever noticed that nights tend to be warmer when the weather is humid or cloudy than when it's dry or clear? If you've ever spent a night in the desert or up in the mountains, you may have noticed that the air cooled off really fast after the sun set.

To understand how the earth's atmosphere affects local air temperatures, you need to know a little bit about the stuff that makes up air. Earth's atmosphere is mostly nitrogen gas and oxygen gas. In a gallon bucket filled with air, you have more than three quarts of nitrogen and a bit less than one quart of oxygen. The spoonful or so left over contains trace amounts of argon, carbon dioxide, neon, helium, methane, and a number of other gases, including water

Layers in the Atmosphere

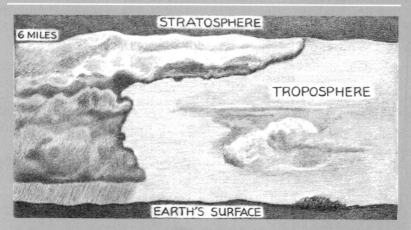

STRATOSPHERE

6 MILES

TROPOSPHERE

EARTH'S SURFACE

I n the earth's atmosphere, layers of air float on top of one another like oil floats on top of vinegar in salad dressing. Just as oil is less dense than vinegar, the upper layers of the atmosphere are usually less dense than the ones above them.

The *troposphere* is the layer forming the lowest six miles or so of the atmosphere. The name comes from the Greek word *tropos,* which means "turning." Air in this layer does a lot of turning over, rising to form roiling clouds, sinking to create gusts that roar down alpine canyons. Within this layer, the temperature usually decreases at about 3.5°F per 1,000 feet, as you travel upward from the surface.

Above the troposphere lies the *stratosphere,* which extends from about six miles to about 30 miles above the surface. Cross-country jet flights usually travel in this layer, since it's usually calm compared to the layer below and has a lower density, which reduces the drag on planes. Temperatures hold steady or increase as you go up in the stratosphere. That's because this layer has a relatively large amount of the gas called ozone, which is good at absorbing ultraviolet (UV) radiation from the sun. As UV radiation is absorbed, the stratosphere warms up.

vapor. Though the atmosphere contains relatively little of these gases, some of them have profound effects on weather and climate.

The sunlight that warms the earth is one form of electromagnetic radiation. (Some other forms are radio waves and X rays.) The gas molecules in the atmosphere can absorb electromagnetic radiation. This causes them to move faster, a change we feel as an increase in air temperature. But gases are selective absorbers—they absorb some forms of electromagnetic radiation very well, some a little bit, and others not at all.

Most of the gases and other stuff in the atmosphere aren't very efficient at absorbing energy from sunshine. Of all the solar energy that hits the top of the earth's atmosphere, about 30 percent is reflected away by clouds, gases, and the earth's surface; this energy contributes nothing to warming our planet. About 19 percent is absorbed by clouds, dust, ozone, and water vapor in the atmosphere. The remaining 51 percent is absorbed by the surface of the earth.

Most of the energy that is absorbed by the water and land on the earth's surface is reradiated later as infrared radiation, another form of electromagnetic radiation. You can't see infrared radiation, but you can feel it as radiated heat.

This change in form—from visible light to infrared radiation—helps keep the earth warm. Carbon dioxide and water vapor—two gases that make up a tiny fraction of the gases in the atmosphere—are very good at absorbing infrared radiation, which warms them up. These gases are known as *greenhouse gases* because, like the glass walls in a greenhouse, they help to slow down the loss of heat. The role of these naturally occurring gases in keeping the earth warm is called the *greenhouse effect.* (For information on how human activity has enhanced the greenhouse effect, see page 133.)

On a sunny summer day in the city, you can feel infrared radiation as heat coming off the pavement. The warming you feel is the result of solar energy being passed from the ground to you. The ground can be pretty slow about releasing this energy, so you can often feel this warming well after the sun has set.

The ground warms the air just above it. As the air gets warmer, it starts to emit more infrared radiation itself. Some of this infrared radiation from the air heads upward, but much of it is sent right back down to the ground. The ground absorbs this radiation, warms up a little, and radiates more infrared radiation back upward into the air. This back-and-forth process can go on quite a few times before the energy eventually escapes into outer space as infrared radiation. The process is a bit like bouncing a Superball between the floor and the ceiling of a room with an open skylight. The ball, like the infrared radiation, bounces up and down in a pretty crazy way. After many bounces, the ball escapes out the skylight.

All over the earth, surfaces are sending infrared radiation into the air. The hot surfaces are sending lots of infrared radiation, while the cold ones are sending out much less. The atmosphere above all these surfaces slows down the escape of infrared radiation into space by forcing the infrared radiation to do a lot of bouncing around in the air before leaving earth.

The atmosphere, with its greenhouse gases, keeps the world much warmer than it would be without an atmosphere. How much warmer? Well, earth's surface temperature averages about 60°F. Scientists estimate that without the atmosphere, this temperature would be about 0°F, or 60°F colder. This would make earth much more like the moon or Mars, where the atmospheres are much thinner and where temperature extremes have created very inhospitable environments.

So what does all this have to do with hot, humid nights and cold, desert nights, and temperature extremes in the mountains? One of the gases that's best at absorbing infrared radiation is water vapor. Water vapor absorbs about five times more of the infrared radiation from the ground than all the other gases in the atmosphere combined.

On a summer night in a Gulf Coast city, such as Houston, the ground has absorbed a lot of energy from the day's sunshine, and the air has plenty of water vapor collected from the warm gulf waters. When the ground releases infrared radiation, the water vapor absorbs some of it and sends that back downward, keeping that warmth from escaping. The same principle explains why cloudy nights are warmer than clear ones. The water droplets and ice in clouds, along with the humid air in and around them, helps slow down the loss of warmth from the ground.

In desert and mountain air, by comparison, water vapor is in short supply. When the ground releases infrared radiation, there's little to slow the loss of that energy into space. At night, there's no incoming solar energy to replace the energy lost to space, and so the desert and mountains cool off quickly.

Cold Mountain Nights

The thickness of the atmosphere over your head depends partly on your elevation. Gravity pulls air down toward the earth's surface. The air at sea level is compressed by the weight of all the air above it. As you go up in altitude, there is less air above you and therefore less air pressing down on you. Another way of saying this is that the air is under more pressure near the earth's surface and under less pressure at higher altitudes.

The result is an atmosphere in which the density of the gas molecules—how closely they are packed—changes as you travel up into the atmosphere. There are more molecules in a gallon of air when you are at sea level than there are when you are at the top of a mountain.

A breath of air taken in Denver will only deliver 82 percent of the oxygen you would get from the same breath taken at sea level.

If you've ever traveled from sea level to high altitudes and then tried to exercise, you've experienced this change in air density or pressure. If you're used to living at sea level and you go for a jog in Denver, you may find yourself gasping for air. Your body is adapted to sea level, where it gets more oxygen per lungful of air. Denver is about a mile high, and the density of the air there is about 18 percent less than it is at sea level. So each breath brings in only 82 percent of the oxygen that you are accustomed to inhaling at sea level.

Elevation changes can also have a dramatic effect on an area's climate. Consider, for example, Mount Rainier, near Seattle, Washington, which rises almost three miles into the atmosphere. At Longmire Ranger Station, at an altitude of about half a mile, the average high temperature for July is 75°F; the average low is 47°F. At Paradise Ranger Station, at an elevation of just over a mile, the average high temperature in July is 64°F and the average low is 44°F.

You might think that a change of half a mile wouldn't make any difference—after all, the earth's atmosphere is hundreds of miles thick. But keep in mind that the air in the lower altitudes is much denser than the air above it. In fact, half of all of the atmosphere's mass lies within just 3½ miles of the earth's surface. So although the lowest few miles are just a small fraction of the atmosphere's total thickness, they contain more than half of the water vapor, carbon dioxide, and other greenhouse gases that slow the loss of infrared radiation to space. Since the atmosphere is thinner above mountains, nighttime temperatures drop more quickly than they do at sea level.

That's one of the reasons that mountaintops differ in temperature from nearby lowlands, but it's not the only factor. As air rises from low altitudes to high altitudes, it expands and cools off. This process of expansion and cooling is responsible for the clouds that often form at mountaintops. (See page 28.)

The Earth's Energy Budget

Light from the sun helps warm the earth, and infrared radiation lost to space helps cool it down. Together, this input and output of energy helps determine the earth's overall temperature. The balancing act of inputs and outputs determines what scientists call "the earth's energy budget." If the inputs equal the outputs, the earth's overall temperature will remain constant. But if they're not equal, the earth will warm up or cool down.

Over the centuries, the earth's energy budget has shifted in one direction or another. Volcanic eruptions can send dust clouds high into the atmosphere. By reflecting sunlight back into space and decreasing the energy that reaches the earth's surface, this dust can decrease the incoming energy and cause the earth to cool. In 1815, Mount Tamboro in Indonesia erupted, sending fine dust more than 15 miles into the atmosphere. During the following year, sometimes known as the "year without a summer," New England states had heavy snowfalls and Vermont lakes had inch-thick ice in June. Crops failed, not only in the United States, but also in Canada and Europe.

Elevation changes can have a dramatic effect on an area's climate. Since the atmosphere is thinner above mountains, nighttime temperatures drop more quickly than they do at sea level.

More recently, following the 1991 eruption of Mount Pinatubo in the Philippines, the solar energy reflected from the atmosphere increased by several percent, an effect that lasted for many months. Satellite measurements indicated that over much of the earth, the lower atmosphere cooled by an average of 1°F. This may have been a factor in the unusually cool summer experienced from the Rockies into the northeastern U.S. in 1992.

Reducing incoming solar energy can cause the earth to cool down. Similarly, reducing the rate at which energy leaves the earth can increase the planet's temperature. Carbon dioxide is one of the gases that absorbs infrared radiation, preventing it from radiating out into space. For more than a century, the amount of carbon dioxide in the atmosphere has been rising steadily as people have burned more and more carbon-based fuels, such as coal, oil, gasoline, and natural gas. Adding more carbon dioxide to the atmosphere may lead to a global warming of the air near the earth's surface, a human enhancement of the greenhouse effect resulting from naturally occurring carbon dioxide and water vapor. (We'll talk about this more in chapter 6.)

Microclimates in Your Yard

I f you walk outside on a cold winter morning, you may find frost on your car windows, but not on the tires or sidewalk. As spring comes on and the snow begins to melt, some patches of snow around your house can last much longer than others. In the summer, as you dig in your yard, you may find places where the soil is warm and dry, and other places where it's cool and damp. These little changes from place to place are all created by different surface conditions.

The windows of your car frost up because the glass radiates infrared radiation quickly and gets cold enough for water vapor to condense directly as frost, while the tires, which remain warmer, stay frost-free. The patches of ground that stay snowy longest are the ones well shaded from the sun. These are often patches that face north, away from the sun. This tilt away from the sun is often the reason why soil on the north side of a building, or on the north-facing side of a canyon, is much damper than on the south side.

The small-scale changes of the surface help define what's known as microclimates—the little climates within bigger ones. But even the bigger climates are closely controlled by what's going on at the surface. The moderating effects of watery surfaces, for instance, are critical in determining both the climate around a small pond and the climate of western Europe, located on the downwind side of the North Atlantic.

Heat on the Move

The way that sunlight is absorbed and infrared radiation is emitted by the earth makes some places hot and some places cold. This has enormous consequences for weather on our planet.

Heat moves from hot places to cold ones. If you put an ice cube on a hot sidewalk, heat flows from the sidewalk to the ice cube, the ice cube warms up and melts, and the sidewalk under the ice cube cools down.

In the atmosphere, the transfer of heat from warm regions to cool regions is constantly under way. The earth's surface, which is warmed by the sun, warms the air above it. The warm air rises upward into the cooler regions of the troposphere (see page 18), and cooler air from above sinks to the surface to replace the rising air. This process is known as *convection,* and understanding it is key to understanding the weather. As we'll see in chapters 2 and 4, this thermally driven convection is critical in producing the earth's major winds and storms.

On the global scale, the temperature differences between hot places, such as the tropics, and cold places, such as the Poles, drive the winds and ocean currents. The winds and currents transport heat from hot regions to cold ones, helping to make the tropics cooler and the Poles warmer than they'd otherwise be. This process gives the in-between regions, the midlatitudes, a mixture of cold, cool, warm, and hot weather. In these regions, the clash between warm air from the tropics and cold air from the Poles creates blizzards, thunderstorms, tornadoes, and other severe weather conditions.

Meteorologists can often figure out the immediate cause of such weather events. But nature usually doesn't leave enough clues for them to trace the disturbances back for more than a few links in the chain of events. Temperature differences produce changes in the atmosphere, generating winds, rain, and snow. And once these weather processes get started, they can alter the local temperature differences from which they arose, leading to new temperature differences and new wind, rain, and snow patterns. These complex interactions are one of the main reasons that predicting the weather is so difficult.

2Wind

Air on the move

WHEN MARY MILLER BEGAN BICYCLE COMMUTING to the Exploratorium along San Francisco's waterfront, she didn't know the weather was going to be so cooperative. Oh sure, she knew that the climate was good for cycling. It's not particularly rainy and the daytime temperatures rarely drop below 50°F or rise above 80.

In the morning, the air is usually calm, but on most afternoons, a good stiff breeze picks up. This afternoon breeze, called a "sea breeze" because it flows from the sea to the land, brings cool air from the Pacific Ocean west of the museum and blows it across the bay toward the warm inland valleys to the east. For Mary, the sea breeze gives a welcome push on her back as she peddles to the train station on the eastern edge of the city.

Perhaps you've also noticed daily or seasonal wind patterns where you live and work. Winds often sweep across valley floors and up the sides of mountains or hills during the day. If you live in an inland city, you may have noticed that winds sometimes blow from outlying rural areas into cities in the summer. In Chicago and other communities near large lakes, residents may experience the equivalent of an afternoon sea breeze, also called an onshore or "lake breeze" because it blows onshore from the water to the land. At night, the wind often switches to an offshore or "land breeze," which blows from the land to the water.

The wind is like the air, only pushier.

—Test response by fifth-grade student

All these weather patterns are the result of a very basic physical phenomenon: Warm air rises and cool air sinks. As you know from the last chapter, different surfaces on the earth and the air above those surfaces warm up and cool down at different rates. The resulting temperature differences power the wind.

Consider the circumstances that create Mary's biking weather. At 7:00 A.M. when she's riding westward into work, the land surfaces around San Francisco Bay have had all night to cool off and they're about the same temperature as the nearby ocean, somewhere in the 50s. At that point, there's usually little or no wind.

As the sun rises higher, it warms the inland valleys faster than it warms the ocean. (See pages 10–12 to understand why.) As the temperature of the land rises, the air just above it gets warmer. When air heats up, the individual molecules start moving faster and bouncing off each other more often and harder. The molecules spread apart as they bounce off each other, and the air expands. This expanding warm air has a lower density than the surrounding air and it begins to rise, like oil floating to the top of an oil-and-vinegar salad dressing.

Cool under Pressure

Next time you drive up into the mountains or take a plane ride, take along an empty, airtight plastic bottle. Close it before you ascend. When you get to the top of the mountain or when your plane reaches cruising altitude, you may notice that your bottle has ballooned outward.

Air pressure decreases with altitude, and the high pressure air that you trapped inside the bottle before you started out is easily able to push outward against the low pressure, high-altitude air surrounding the bottle. If you open the bottle at the mountaintop or in the middle of your flight, the bottle will fill with low pressure, high-altitude air. Seal it again; when you descend, the bottle will collapse as the higher pressure air around it pushes in.

Squeeze a sealed bottle filled with air, and you'll feel it resisting your inward pressure. To compress the air in the bottle, you have to exert yourself and use up some energy. Similarly, to expand the air in the bottle, you'd have to use energy to push against the outside air that's squeezing in on it.

When air rises in the atmosphere, it expands by pushing away the surrounding air. This takes energy from the air, which causes its temperature to drop. So ascending air cools down by what's called *expansion cooling.*

Likewise, when air descends, the surrounding air compresses it, giving energy to the descending air. So the descending air warms up by what's called *compression warming.*

Whenever air rises or falls, it cools off or warms up. This is a major reason why the temperature decreases with altitude in the troposphere.

Air forced up over a mountain is cooled by expansion. This cooling can condense the water vapor in the air, forming the clouds that you often see surrounding mountaintops. Air flowing down a mountain warms up and can form hot, dry winds, such as Chinooks and Santa Anas.

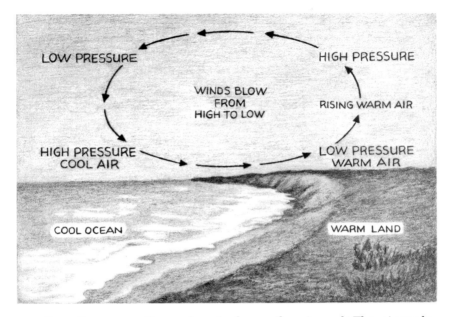

LOW PRESSURE

HIGH PRESSURE

WINDS BLOW
FROM
HIGH TO LOW

RISING WARM AIR

HIGH PRESSURE
COOL AIR

LOW PRESSURE
WARM AIR

COOL OCEAN

WARM LAND

Over the ocean, the cool water keeps the air cool. The air molecules are moving more slowly and bouncing off each other less often and with less force. In this cool air, the molecules pack together more closely, making the air denser and heavier than warm air. So the cool air stays close to the ocean's surface. If you've ever gone into an attic on a sunny afternoon, you've probably noticed that warm air rises and cool air sinks—it can be tens of degrees hotter up in the attic than in the rooms below.

As warm air rises over the valleys east of San Francisco, it leaves behind an area where there are fewer air molecules per cubic foot. A meteorologist would call this a *low pressure area,* or a *low,* because the warming of the air and the decrease in air density lowers the air pressure in this area. Just above the ocean's surface, the cooling influence of the water creates an area where there are more air molecules per cubic foot. A meteorologist would call this a *high pressure area,* or a *high,* because the cooling of the air and the increase in air density raise the air pressure there.

Air pressure refers to the ability of air to push out against its surroundings. A pressure difference within the atmosphere pushes air from the high pressure area into the low pressure area, creating wind. The space left behind by the rising air over the warm land leaves room for air to rush in from the higher pressure area over the ocean. By mid-afternoon in San Francisco, the air is often blowing at a brisk clip from the cool Pacific toward the inland valleys, creating the eastward tailwind that speeds Mary and her bicycle back home.

Predicting Wind

Knowing the locations of higher and lower pressure areas is a key to understanding the weather, which is why TV meteorologists almost always mention air pressure (sometimes called *barometric pressure*) in their broadcasts. On weather maps, high pressure areas are labeled "H" and low pressure areas are labeled "L." These terms are always relative to nearby areas at about the same altitude; there are no absolute standards for what makes a high and what makes a low pressure area. The high and low designations only indicate that there's a pressure difference between two or more places.

Maybe you've noticed when a TV meteorologist mentions that the pressure is falling, she may also predict that a storm is coming. In general, low pressure areas are associated with unstable or rapidly changing conditions. That's because air moving toward lower pressure areas often brings clouds and precipitation along with it. The rising air over a low pressure area can also create clouds and storms.

High pressure areas are more stable and are usually associated with clear weather. Air moves downward over these areas, which makes it difficult for clouds to form.

In late summer and early fall, high pressure areas can create clear skies and relatively warm weather in the northeastern United States. This pleasant weather, often called "Indian summer," usually follows a period of cool or cold early fall weather. The high develops because the land cools faster than the nearby ocean, which causes air to descend. As the air descends, it warms up (see page 28). The result is a period of summerlike weather that typically lasts a few days to a week. The name Indian summer

Reporters sometimes blame the air pollution in my city on a "temperature inversion." What's that?

Normally, the air near the earth's surface is warmer than air higher in the troposphere. This warm air rises and is replaced by air descending from above. In cities where air pollution is a problem, this air circulation helps move the dirty air produced at street level upward, and bring cleaner air downward.

Sometimes, however, the air near the ground is cooler than the air above. Since this is the inverse of the normal temperature pattern, it's called a *temperature inversion*. Warm air above cool air is a stable arrangement, so there's little or no mixing of air at street level with the air above. So pollution collects near the ground, producing choking smog.

comes from New England; in some parts of Europe, the same phenomenon is called "Old Wives' summer" or "halcyon days." Halcyon comes from *alkuon,* a mythical Greek bird that could calm winter winds and waves.

Knowing where areas of higher and lower pressure are can also help you understand and predict the direction and timing of local winds. You can even calculate wind speed if you know the difference in pressure and the distance between the higher and lower pressure spots. Meteorologists use this calculation every day to describe and predict winds. The gist of their calculation is simple: The strength of the wind depends on the pressure difference between two points and the distance between the points. For two points that are a given distance apart, a larger pressure difference will produce a stronger wind. For a given pressure difference between two points, a shorter distance will produce a stronger wind.

The change in pressure between two points, divided by the distance between them, is called the *pressure gradient*. If you look at a pressure map, it looks very similar to a topographic map. Just as the contour lines on a topographic map connect points that are at the same altitude, the lines on a pressure map connect points that are at the same atmospheric pressure. On a topographic map, the closer the lines are to each other, the more steeply the land slopes. The same relationship is indicated on pressure maps: The closer the lines are, the steeper the pressure gradient. A high pressure area is like the top of a hill, while a low pressure area is like the bottom of a valley. Water flows from the top of a hill to the valley, flowing fastest where the slope is steepest. Air tends to flow from the high pressure area to the low pressure area, blowing fastest where the pressure gradient is steepest. At the center of the high pressure area and the center of the low pressure area, winds are weak because the pressure gradients there are small.

A high pressure area is like the top of a hill.

In the valleys thirty miles east of San Francisco, the daytime temperatures in summer can soar to over 100°F, but the nearby ocean stays in the 50s. This extreme temperature differential creates a steep pressure gradient, which drives onshore breezes that can blow at speeds of 40 miles per hour or greater, giving Mary the equivalent of a downhill ride after work.

Wind and pressure gradients can usually be traced back to differences in the earth's surface temperature. If you live in a valley surrounded by hills or mountains, you've probably noticed that the breeze often picks up in the afternoon. Early in the day, the mountainsides that face the sun warm up faster than the shaded valley floor. As warm air rises up from the mountain, it's replaced by cooler air from the valley, creating a breeze that blows up the mountain. You can often see hawks soaring on these *thermal updrafts*, created from warm air rising up the sides of a mountain.

Wind can also develop when an area cools down faster than its surroundings. If you live on the coast, you may have noticed that sea breezes blow from a relatively cool ocean or lake to the warm land during the day. At night, the wind may reverse direction and begin blowing from the land to the water. Called *land breezes*, these reverse winds develop because the land cools off much faster than

Does a windchill of -17°F really mean that it feels as if it's -17°F outside?

Wind makes you feel colder because it increases the loss of heat from your body. The faster the wind blows, the faster your body loses heat and the colder you feel. The *windchill,* or *windchill equivalent temperature* (WET) as it's known to meteorologists, is an estimate of how the actual air temperature and the wind combine to cool unprotected human skin.

On a day with an air temperature of 32°F and winds of 12 mph, the WET would be about -17°F. That means your bare skin would lose heat at about the same rate it would if you were standing in -17°F still air. Your skin won't actually cool to -17°F; it will only get as cold as the actual air temperature of 32°F. But all else being equal, you'll feel about as comfortable as you would if the air were -17°F and still.

The WET has some limitations. It doesn't take into account how your body might be warmed by sunlight or how your heat output and body temperature change as you exercise. But the WET can still be a good estimate of how wind and air temperature will affect you.

the water. A higher pressure develops over the cool land and air flows toward the lower pressure over the water. The breeze that blows up a mountain face during the day also switches at night. The bare rocks and high altitude help cool the mountain surfaces faster than the valley below, so winds blow downslope after sunset.

A caution about predicting wind patterns: They can change on you from day to day and season to season. Sea breezes, land breezes, and other local winds may be more obvious in summer than winter. The local pressure gradients that form between nearby cool and warm areas can be overwhelmed by stronger gradients that develop in response to larger-scale temperature differences. In winter, for example, major winds are often determined by strong high and low pressure systems that extend across thousands of miles.

Upwinds and Downwinds

Wherever warm air is rising from below or cool air is sinking from above, vertical winds are blowing up or down in the troposphere, the bottom layer of the atmosphere where most of our wind and weather are generated. Together, these vertical winds and the winds that blow horizontally complete a circuit.

It works like this. As air warms up, it expands, and the same number of gas molecules occupy more space. This low-density air rises, pushed upward by the air around it, leaving a low pressure area behind. Higher in the atmosphere, the air gets thinner—that is, its density decreases. The rising air is pushed up as long as it is surrounded by air that has a higher density. When the rising air reaches the point where its density matches that of the surrounding air, the surrounding air stops pushing it upward, but momentum carries the rising air farther, until it's at a higher density than the air around it.

This process creates a column of air, extending up from the earth's surface, with a low pressure area at the bottom. The accumulation of air at the top of the column causes the pressure there to be higher than the pressure of the surrounding air at the same altitude. In general, the warmer the rising air is, compared with the surrounding air, the higher the air will rise and the taller the air column will be. For local sea breeze circulations, the top is about a mile above the surface. For air warmed over a tropical ocean and raised up in a hurricane, the top of the column might be 10 or more miles high.

The opposite happens over a cool surface, like the northern Pacific Ocean in the summer, or Siberia in the winter. As air molecules are cooled, they sink and accumulate just above the surface. As air sinks within the air column, the column gets shorter and an area of low pressure develops at the top of the column. Air tends to flow from the high pressure area at the top of the column of rising air into the low pressure area at the top of the column of sinking air. This flow creates winds high in the atmosphere that blow in the opposite direction to the surface winds.

With all this warm air rising and cool air sinking, you might be thinking the upper troposphere is a hot place and the lower troposphere is cold. That's not the case, as you know if you've ever gone to the top of a mountain. As warm air rises in the troposphere, it expands and cools off. As air compresses and sinks, it warms up. This expansion cooling and compression warming helps keep the upper air cooler and the surface air warmer than what you might expect.

Vertical winds are usually weaker than surface winds, but they have profound effects on the weather. They make horizontal winds possible, and they are essential for storm formation (as you'll see in chapter 4).

Vertical winds are also responsible for a common summer weather pattern—a bright clear morning that becomes a cloudy afternoon. What causes the change? As the sun warms the surface in the morning, the air starts to rise. As it does, it expands and cools. Water vapor contained in the air condenses into clouds, which can spawn afternoon rain and thunderstorms. The day may end with a vivid red sunset as the clouds begin to move out. By morning, the skies are often clear again and the cycle starts over.

Winds that blow from as far away as the Gulf of Mexico can travel across the plains until they reach the eastern slope of the Rockies.

Something similar can happen on the windward side of a mountain. When horizontal winds run into the mountains, they're forced uphill by the land and by the air following behind. These are called *upslope winds.* Local forecasters in Denver are usually breaking bad news when they tell residents that upslope weather is on the way. Winds that blow from as far away as the Gulf of Mexico can travel across the plains until they reach the eastern slope of the Rockies. As the warm, humid air from the south starts climbing up the mountain range, it cools and condenses to form clouds and rain or snow.

Upslope winds can happen in any mountain range and account for mountains usually having more snow or rain than nearby lower elevations. If you drive up into the mountains on summer vacation, you may have noticed that they're often clouded over while the valleys below are clear. You can also see this on a weather satellite image: There's often a band of puffy, white clouds located directly over the mountains.

In the Cascades and the Sierra Nevada, winds from the Pacific travel up the western slopes, causing expansion cooling (see page 28). Water vapor condenses from the cool air, forming clouds on most summer afternoons. Many an inexperienced backpacker has been caught in an unexpected downpour after being lulled into a false sense of security by a clear, sunny morning. To avoid being caught by surprise, Paul Doherty, Exploratorium physicist and experienced mountaineer, follows this rule of thumb: If puffy cumulus clouds build before noon so that they are higher than they are wide, he worries about midafternoon thundershowers. Blue skies in the morning are nice, he says, but in the mountains they are no guarantee of a dry afternoon.

Upslope winds in the mountains can bring clouds, rain, and snow, but the winds that flow down a mountain usually bring clear skies and warmth. Forecasters in the Great Plains often speak of the downslope flow as being responsible for their pleasant weather. As air flows up and over the western slopes of the Rockies, it drops most of its moisture on the western slopes and tops of the mountains. The drier air continues its trip across and down the eastern slopes. As it descends, the air compresses and warms up. By the time it hits the plains, the air is very dry and warm enough to start melting snow. These downslope winds off the Rockies are known as *Chinook* winds, which literally means "snow eater."

The Santa Ana winds are downslope winds that flow westward off the mountains of southern California. They are common in the fall, when the desert to the east of the mountains begins to cool and a higher pressure develops there. Dry desert air blows from this high pressure area toward the lower pressure off the Pacific coast. These winds can feel like a furnace blast as they flow down the mountains toward the Los Angeles basin, drying out the vegetation as they go. Tom recalls that when he lived in San Diego, he could look north along the beach and see when the Santa Ana winds

How fast is the wind blowing?

I n 1805, a British naval officer named Francis Beaufort classified winds at sea by observing their effects on the water. You can use this modified version of Admiral Beaufort's scale to estimate wind speed.

Beaufort Scale

Force 0	less than 1 mph	Calm. Smoke rises straight up.
Force 1	1–3 mph	Light air. Weather vanes are still, but smoke drifts.
Force 2	4–7 mph	Light breeze. You can feel the wind on your face and hear leaves rustle.
Force 3	8–12 mph	Gentle breeze. Leaves are in constant motion. Flags wave.
Force 4	13–18 mph	Moderate wind. Raises dust and loose paper. Moves small branches. Flags flap.
Force 5	19–24 mph	Fresh wind. Small trees begin to sway. Waves form on lakes.
Force 6	25–31 mph	Strong wind. Large branches move. Umbrellas are hard to use.
Force 7	32–38 mph	Near gale. Whole trees bend and sway. It's hard to walk. Flags fly straight out from the pole.
Force 8	39–46 mph	Gale. Wind breaks twigs off trees. It's very hard to walk.
Force 9	47–54 mph	Strong gale. Tree branches snap. Damage to signs, awnings, antennas.
Force 10	55–63 mph	Storm. Trees are uprooted. Damage to buildings.
Force 11	64–72 mph	Violent storm. Widespread damage.
Force 12	over 73 mph	Hurricane. Extreme damage.

were blowing through Los Angeles. The signal was a long brown river of air heading out to sea—the polluted air of Los Angeles being swept offshore by the Santa Anas.

Santa Ana winds, like other winds that flow over rough terrain, can be very gusty. As the wind squeezes through canyons and up and down mountains, friction helps create eddies, or swirling winds. These eddies are very much like the turbulent eddies that form behind rocks in a fast-moving stream. As the wind hits obstacles in its path, it alternately speeds up, stops, and even changes direction like the eddies in a rocky stream. Together, these gusty, swirling winds and dry air can create dangerous fire conditions. Some of the worst wildfires in southern California have been whipped into a fury by the Santa Anas.

There was a desert wind blowing that night. It was one of those hot, dry Santa Anas that come down through the mountain passes and curl your hair and make your nerves jump and your skin itch. On nights like that, every booze party ends in a fight. Meek little wives feel the edge of the carving knife and study their husbands' necks. Anything can happen.

—Raymond Chandler,
Red Wind

Turbulence can develop wherever flowing air runs into or past obstacles, including other air. For instance, as air blows up and over mountains, it may meet another layer of air moving in a different direction or at a different speed above the mountain. As the two layers move past each other, swirls and eddies will develop along their boundaries. Planes flying above the Rockies often run into these invisible swirling winds, an example of what's called *clear air turbulence*. Turbulence in the atmosphere accounts for gusty winds and is an important factor in stormy weather.

Long-Distance Winds

The pressure gradients that create local wind patterns also dictate global patterns. Winds can operate over vast distances, both near the surface and higher in the troposphere. The surface winds that are part of the Asian monsoons flow into southern and eastern Asia from across thousands of miles of ocean during the summer, and reverse to flow in the opposite direction during the winter. The summer monsoon winds begin when the land warms up under the fierce summer sun, creating a strong low pressure area over Pakistan and northern India. Moist marine air flows in from the higher pressure areas over the southern Indian Ocean and the Pacific Ocean.

With the summer monsoon winds come deluges. Moisture carried by monsoon winds and the resulting rain is vital to farmers from Pakistan to Vietnam and northward to northern China, but these rains can also cause devastating floods.

In winter, the monsoon winds switch direction. The air over Siberia and China cools dramatically, producing a strong high pressure system. Cold, dry air flows southward from this high toward lower pressure areas in the tropics and the Southern Hemisphere, where summer is developing. The winter monsoon winds bring extended periods of dry weather over southern and eastern Asia.

A weak version of the Asian monsoons occurs in other parts of the world, including the southwestern U.S. and Mexico. In the summer, rising air over the hot interior creates lower pressures, which bring in moist surface air from the Gulf of Mexico and the eastern Pacific. This causes summer rains, and sometimes thunderstorms and flash floods in the deserts. The monsoon weather pattern in the American Southwest is not as dramatic as it is in Asia because the temperature and pressure differences are not as extreme, and because extensive mountain ranges in Mexico prevent much of the moist marine air from reaching the interior.

Coriolis and Cannonballs

In the early 1800s, French scientist Gustave-Gaspard Coriolis was asked by the army to explain why its new long-range cannons weren't hitting their targets. Instead, the cannonballs were consistently landing too far to the right. Coriolis determined that the earth's spin was causing the cannonballs' paths to curve to the right. He worked out the math and found that what we now call Coriolis effects deflect objects from the paths they'd follow if the earth was not spinning—to the right of their paths in the Northern Hemisphere and to the left in the Southern Hemisphere. These effects are important for determining the paths of objects that travel long distances or for long periods over the earth's surface, such as planes making transcontinental flights and rockets and space shuttles traveling around the earth. They also need to be taken into account in forecasting the weather and the flow of the ocean.

You may have heard that Coriolis effects make water draining from bathtubs and sinks swirl counterclockwise in the Northern Hemisphere and clockwise in the Southern Hemisphere. That's a bit of physics folklore. At the very small scale of a sink or bathtub, Coriolis effects have a negligible influence on the direction of the vortex that forms when the water drains. The direction in which the water swirls is primarily determined by the shape of the basin, the residual motions of the water (from when the basin was filled), and other small-scale factors.

Coriolis Effects

The monsoon winds are in many ways like very long-distance sea and land breezes. But winds that travel thousands of miles do not beat a straight path from a high pressure area to a low pressure area. Instead, the winds curve. For example, in July, air from a high pressure area east of Madagascar (south of the equator) heads for

South of the equator, the path of the Asian summer monsoon winds curves to the left; north of the equator, it curves to the right. That's because these long-distance winds are traveling over the surface of a spinning planet.

Low pressure

E q u a t o r

High pressure

the low pressure area over Pakistan and India. But instead of traveling due north, it first curves to the northwest toward East Africa. As it approaches the equator off the coast of Kenya, it starts to turn northward. After entering the Northern Hemisphere, it curves northeastward, crosses the Arabian Sea, and blows into India.

Why such a roundabout path? Because the wind is moving over a spinning sphere. When you're standing on the earth, you feel like you're standing still, but you're actually spinning with the earth. Exactly how fast you're moving depends on where on the planet you're standing. If you're standing at 45°N latitude, the spinning earth is carrying you and the atmosphere along at about 730 mph. If you're closer to the equator, at 30°N latitude, you move a greater distance in the same amount of time, traveling at about 870 mph.

To understand why the wind curves, consider air that's flowing due south, from a high at about 45°N latitude to a low at 30°N latitude. At 45°N latitude, before it begins to move southward, the air is traveling with the earth at 730 mph. As the air heads south, it lags behind the earth's surface, which, at 30°N latitude, is traveling at 870 mph. If you were standing on the spinning earth and watching this air, you would see it curving off to the west as it heads south. If you were looking in the direction that the wind is blowing, you would see the air curve to the right.

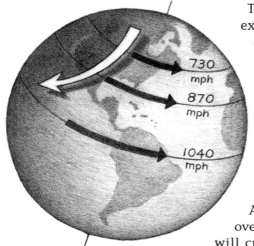

This curving motion is one example of what are called *Coriolis effects,* named after the French scientist who first described them. (See page 40.) We've explained why a wind blowing southward in the Northern Hemisphere curves to its right. But it doesn't matter whether a wind is blowing north, south, east, or west. As long as the air is moving over a spinning planet, the wind will curve. Though the detailed explanation of why the wind curves depends on the direction of the wind, the reasons always relate to the spin of the earth.

Winds in the Northern Hemisphere curve to their right. In the Southern Hemisphere, winds curve to their left. That makes the summer monsoon blowing from Madagascar curve first to its left (toward East Africa). Then, after entering the Northern Hemisphere, it curves to its right (to the northeast and India).

You can see Coriolis effects on your local weather. Pay attention to the satellite movie portion of a TV weather report. Look for the sickle-shaped bands of clouds in hurricanes and larger storms and notice that the clouds spiral around as they move. That spiraling motion shows Coriolis effects.

Sailing Ships and Trade Winds

Although Columbus was unaware of it, his discovery of the Americas would not have been possible without Coriolis effects. When Columbus set out from Italy in 1492, he hitched a ride on the trade winds, which blow southwestward from the coasts of Portugal and North Africa to South America and the Caribbean Sea. If the

W A T C H I N G   W E A T H E R

surface winds that Columbus caught flowed in a straight line from the high pressure area off western Europe to the low pressure area at the equator, he would have ended up somewhere off the coast of central Africa. But the Coriolis effects turn those southward winds to their right, so they become the southwestward trade winds that help sailors cross the Atlantic.

Sailors crossing other oceans found similar trade winds that blow out of the higher pressure areas over the subtropics toward lower pressure areas over the hot tropics. In the Northern Hemisphere, the trade winds blow to the southwest. In the Southern Hemisphere, they blow to the northwest. The trades are steady winds that are sustained over thousands of miles because the tropical-to-subtropical pressure gradient that drives them is very extensive and persistent.

As the trade winds from the Southern and Northern Hemispheres meet near the equator, they converge at a low pressure area. There the air rises and the winds weaken. Sailors who dreaded getting stuck there called this region the "doldrums." Meteorologists call this

Naming Winds

The naming of winds is a confusing meteorological convention. A wind that is traveling from west to east (eastward) is called a westerly wind. Meanwhile, a wind traveling from east to west (westward) is an easterly wind, and so on for southerly winds, northerly winds, and other winds.

This way of naming winds focuses on the direction that the wind is coming from. It may seem backwards, but there is logic behind it. Winds bring weather with them, so it's important to look into the wind when trying to predict what the weather will be.

Sailors, on the other hand, are more likely to describe wind traveling from west to east as an eastward wind. If you use the winds to travel, it may be more important to know where the wind will take you, rather than where it is coming from.

region of converging winds and rising air the Intertropical Convergence Zone (ITCZ). The ITCZ often shows up on satellite photos as a band of white clouds running roughly east-west near the equator.

The trade winds are among the most reliable, persistent winds on the planet, and were known to mariners well before the fifteenth century when Columbus sailed across the Atlantic. In the early 1700s, British barrister-turned-meteorologist George Hadley set himself the task of explaining the trade winds that sailors had discovered. Hadley proposed a global pattern of air circulation powered by the strong temperature differentials between the equator and cooler regions to the north and south.

With the addition of Coriolis effects discovered later, Hadley's theory reads like the description of any local wind, except it operates on a global scale. Warm air rises in the tropics and flows at high altitude toward the Poles. Because of the earth's spin, these high winds don't go directly to the Poles but curve off to the east, creating what are called the *upper tropospheric westerlies*. Along the way, the air cools, becomes more dense, and falls back toward the surface.

Hadley proposed that the air descended at the Poles, but actually the air descends in the subtropics, at about 30° north and south latitude. The air doesn't reach the Poles in large part because Coriolis effects deflect it to the east and prolong its poleward journey long enough that it cools before reaching the Poles. The deflected air flows downward in the subtropics, creating high pressure areas at the earth's surface. Air from these highs flows back toward the equator to replace the rising air at the ITCZ, creating the trade winds. Because of the earth's spin and Coriolis effects, the air flowing out of the subtropics does not flow straight toward the equator. Instead, the winds curve off to the west, becoming the northeasterly trades and the southeasterly trades.

Hadley's work was ridiculed at the time. Later, scientists discovered global circulation patterns like the ones Hadley predicted, with circulation cells made up of vertical and horizontal winds. Today, the atmosphere's main tropical-to-subtropical circulation cells are

Hadley Cells

Different regions of the earth have different characteristic winds. The tropics have trade winds, which blow from the east and toward the equator. The midlatitudes have westerlies (winds from the west), and the high latitudes have easterlies (winds from the east). These surface winds are part of large three-dimensional circulations that extend from the surface to the top of the troposphere (five to 10 miles above the surface). One of the best known of these is the *Hadley circulation,* in which air blows toward the equator, upward near the equator, toward the poles at the top of the troposphere, and downward at about 30 degrees north or south latitude.

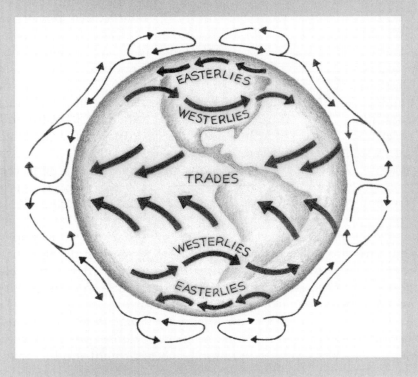

named after Hadley. The Hadley cells originate at the equator and extend to about 30°N latitude in the Northern Hemisphere and 30°S latitude in the Southern Hemisphere. Other cells, called the *polar cells*, extend from about 60° to the Poles in both hemispheres. These cells are driven by temperature differences between the very cold Poles and the warmer, but still cold, subpolar regions.

In between the Hadley and polar cells are the midlatitude winds, which arise from the mixing of warm subtropical air with cool subpolar air. This mixing occurs in large eddies extending over thousands of miles and from the earth's surface to the top of the troposphere. The resulting winds blow mainly from the west, creating the prevailing westerlies of the midlatitudes. In the winter, these westerlies help carry storms out of the North Pacific and across the United States.

Global scale circulation cells dominate the way the earth's atmosphere mixes. They also control the weather and climate in different parts of the world. Where air is rising, clouds and precipitation are likely and the climate tends to be wet. Where air is descending, the climate is generally dry. The world's tropical rain forests are in areas where surface air is rising into the upper troposphere. The world's major deserts are located where upper tropospheric air is descending back to the surface. The Sahara Desert and the tropical rain forests of South America are linked by a portion of the Hadley cell. Air converges at and rises over the low pressure areas, leading to expansion cooling, rain, and rain forests. The air rises to the top of the troposphere and travels northward then northeastward before cooling and descending over the Sahara.

It rarely rains in the desert, even though the surface air temperatures can get very hot, because the descending air keeps air near the surface from rising. As air descends, it compresses and warms up, inhibiting the formation of clouds and rain even when the air contains water vapor. On really warm days, however, the surface air can get hot enough to punch through in places. The rising air cools as it ascends, and water vapor in the air condenses, creating scattered clouds and thunderstorms.

Put Your Back into It

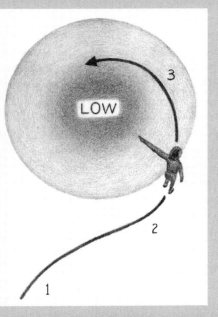

ack in the mid-1800s, Dutch meteorologist Christopher Buys-Ballot discovered that if he stood with his back to the wind, then turned 45 degrees to the right and stuck his left arm straight out, he'd be pointing at the center of the low pressure area that the wind was spiraling into. This trick works in the Northern Hemisphere.

Try it sometime and then compare your result with a newspaper or TV weather map for your area. Since winds can be diverted from a simple path spiraling into a low by such things as cities and mountains, you won't always find the center of the low with this technique, but you'll often come close.

Here's why Buys-Ballot's method works. Air tends to rush into the low pressure area. (See 1 in diagram.) However, as it does, Coriolis effects deflect it to the right. (See 2.) The deflection sends the air counterclockwise around the low. Friction between the air and the surface slows the wind, and weakens the Coriolis effects. This means that the Coriolis effects can cause the air to circle around the low, but can't keep the air from spiraling in toward the center. (See 3.)

Buys-Ballot's method also works in the Southern Hemisphere, except that there you turn to the left and stick your right arm straight out.

Jet Streams

Some of the strongest of all winds are the *jet streams,* rivers of swift-moving air that flow at the top of the troposphere, between about 25,000 and 45,000 feet. These winds were discovered during World War II when high-flying B-29 bomber pilots encountered strong air currents over Japan that slowed their ground speeds to a crawl, even though their instruments clocked air speeds of 300 miles per hour. Pilots today save time and fuel by knowing where the jet streams are blowing. In the midlatitudes, pilots hitch rides on jet streams on their eastbound flights and avoid them when flying west.

Midlatitude jet streams travel in a general west-to-east direction in both hemispheres. When drawn on a map, the jets looks like meandering rivers as they turn north and south. Every meander to the north (called a ridge) and dip to the south (called a trough) affects the weather below it, which is one of the reasons that most meteorologists can't get through a discussion of the weather without mentioning what the jet streams are doing.

The pressure gradients that create jet streams are much like those that propel local breezes, except that they're more extreme. In the northern winter, air rises over Indonesia and sinks over Siberia. At an altitude of about 30,000 feet, the top of the column of rising air over Indonesia is at a much higher pressure than the air at the top of the column of sinking air over Siberia. So at 30,000 feet, air tends to flow from Indonesia toward Siberia. But Coriolis effects cause this wind to curve to its right, creating the eastward-flowing upper tropospheric winter jet stream over East Asia and the North Pacific, the strongest tropospheric jet in the world. This midlatitude winter jet stream has an average peak speed of about 60 mph, but regularly reaches speeds of 150 mph or more.

The winds below the midlatitude jets often travel in roughly the same direction as the jet. The lower level winds tend to be significantly weaker, in part because of weaker pressure gradients at lower levels and in part because of friction with the earth's surface. In this way, these winds are like the currents in a flowing stream.

The water at the top of a stream moves faster than the water at the bottom, but they flow in the same general direction.

When TV weathercasters talk about the jet stream, they are usually talking about the polar jet stream, the jet that has the greatest effect on U.S. weather. This jet sometimes splits into two or more separate flows. There is also a subtropical jet stream at about 25°N latitude, but it usually doesn't affect U.S. weather.

During the Northern Hemisphere's winter, when the temperature differences between the equator and the northern polar regions are most pronounced, the polar jet is at its strongest and often migrates south, bringing cold air and winter storms with it. Since the jet streams play an important role in steering winter storms, we'll return to them in chapter 4.

Jet streams were discovered during World War II by high-flying B-29 pilots.

Whether you're considering a powerful jet stream or a gentle lake breeze, winds and the pressure gradients that generate them are among the most powerful forces that shape our weather. Winds make our planet more livable by cooling off the hot places and warming up the cold places. Winds can also import weather from distant places. In 1983, cold winds swept south out of Canada and into the southeastern United States, causing over a billion dollars in damage to winter vegetable and fruit crops. In 1992, the trade winds carried Hurricane Iniki from its birthplace near Central America toward Hawaii.

Weather forecasters were able to predict the cold snap in 1983 and see Iniki coming in 1992 because they looked upwind and saw what the winds were carrying. You can do the same thing by walking outside and tossing some grass blades or dust in the air to check the direction of the wind. Then look upwind to see what kind of weather the wind is bringing. Another way of looking upwind is to check the satellite pictures on the evening TV news. By looking to the west, you can see whether the jet stream is bringing clear skies or clouds.

3Water

Rain and snow and muggy days

DURING THE SUMMER VACATION MONTHS in San Francisco, it's usually easy to spot first-time tourists. Unprepared for the city's bone-chilling fog and crisp ocean breezes, the visitors shiver in their Bermuda shorts and T-shirts as they walk along the waterfront. This typical West Coast weather pattern was aptly described in a quote attributed to Mark Twain: "The coldest winter I ever spent was a summer in San Francisco."

All that you really need to stave off San Francisco's natural air conditioning is a pair of long pants and a windbreaker. But there's no clothing remedy that can temper the weather for a Californian traveling to steamy Norfolk, Virginia, in July. Stepping off a climate-controlled plane onto the tarmac in Virginia can be an immersion experience: One minute you're cool and dry and the next your shirt is plastered to your back and you feel like you're walking through warm, wet molasses. Although both Norfolk and San Francisco are at about the same latitude and each has an ocean next door, their summer weather is vastly different. The average high temperature in San Francisco in July is 68°F, while Norfolk peaks at about 88°F.

Water vapor in the air does much more than make you miserable on muggy days.

In Norfolk, you aren't just hot—you're both hot and sticky. Your body is sweating in an effort to cool down, but not all of that sweat is evaporating. How readily your sweat evaporates (and consequently how cool you stay) relates to humidity, the amount of water vapor in the air. The combination of high temperatures and high humidity can make the summer weather in Norfolk nearly unbearable to San Franciscans.

Water vapor in the air does much more than make you miserable on muggy days. It's an important greenhouse gas, which helps make the earth a hospitable planet for life. (See page 19.) When it condenses from vapor to liquid, atmospheric water turns into dew, fog, clouds, rain, snow, hail, and other forms of precipitation that supply the planet with fresh water. You probably learned in elementary school about water's transformation from gas to liquid to solid, and the movement of these different forms of water around the earth. This continuous process is known as the *water cycle*. The movement of water into, out of, and through the atmosphere is a major factor in producing both global and local weather.

This chapter considers water's journey through the atmosphere— how it gets into the air, where it goes and what it does once it's there, how it comes back out again, and how all those processes affect the weather and how the weather makes you feel.

Sweat, Clouds, and Latent Heat

On a hot day, you cool off by fanning your sweaty face. Meanwhile, off in the distance, thunderclouds grow high in the sky. Both these things happen because of the peculiar properties of water and what happens when it evaporates or condenses.

Each water molecule is made up of a single oxygen atom bound to two hydrogen atoms, an arrangement that looks a little like a Mickey Mouse hat. The hydrogen atoms of each water molecule are attracted to the oxygen atoms of other water molecules. In liquid water and ice, this attraction creates weak bonds that help hold the molecules together.

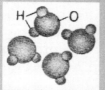

Because of these bonds, it takes a lot of energy to evaporate water—over two million calories to evaporate one gallon. Each year, about 92 quadrillion gallons (that's 92 followed by 15 zeros) of ocean water evaporate. To supply the energy required for all this evaporation, you'd need the energy output of a medium-capacity, 100-megawatt power plant operating continuously for 27 billion years.

Some of the energy needed to evaporate water is drawn from the immediate surroundings—like your hot face—which cool off a bit as a result. That's why sweating helps you cool off.

When molecules of water vapor come back together to form liquid water or ice, this energy, known as *latent heat,* is released and warms the surroundings. When water vapor in rising warm air condenses to form cloud droplets, latent heat warms the air. This, in turn, makes the air less dense so that it rises higher in the atmosphere, which helps the clouds grow taller. Latent heating is essential to the development of thunderstorms and hurricanes, which we'll cover more in chapter 4.

Water Cycle

T he movement of water through the environment is called the water cycle. In its movements, water has a strong effect on weather. For example, water vapor condenses to form clouds, which create shade and precipitation. When liquid water evaporates, it cools the surrounding air, and when water vapor condenses, it warms the surrounding air by releasing latent heat.

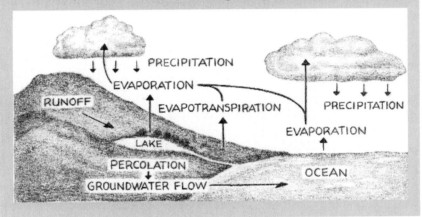

Water in the Air

Water can simultaneously evaporate from and condense into the ocean, lakes, puddles, or other sources of liquid water. The amount of water vapor in the air affects the balance between evaporation and condensation. When the humidity is low, water evaporates more readily than it condenses, and the amount of water vapor in the air increases. As the humidity increases, the balance begins to even out and water vapor condenses at nearly the same rate as liquid water evaporates.

High air temperatures help keep water vapor in the air from condensing back into liquid water. When air warms up, more liquid water can escape into the atmosphere. So warm air can contain more water vapor than cool air. If that warm air cools off, water condenses more readily, turning from invisible vapor into visible liquid. You can see this on a cold morning, when water vapor in your warm exhaled breath condenses into clouds as it's chilled by the cold surrounding air.

The air is saturated when evaporation and condensation are in balance. Water vapor content is often described by the *relative humidity*, which is the ratio of water vapor in the air to the maximum, or saturation, amount for that temperature. In weather reports, relative humidity is usually given as a percentage. A value of 60 percent indicates that the air contains 60 percent of the vapor it could contain at its present temperature. At 100 percent relative humidity, the air is saturated and no net evaporation occurs.

Another measure of humidity is *absolute humidity*, the amount of water vapor per unit of air. The absolute humidity tends to stay fairly constant over the course of a day. For it to change, the amount of water vapor must change—water must evaporate or condense. If the temperature changes but water vapor doesn't enter or leave the atmosphere, the absolute humidity doesn't change.

Relative humidity, on the other hand, depends on temperature, and it changes during the day as the air temperature changes. In the morning, when air temperatures are cool, the relative humidity is typically at its highest level. In the afternoon, as temperatures rise, the relative humidity drops to its lowest level.

Compared with other gases in the air, the actual amount of water vapor is quite low. For example, at 95°F the air contains at most about 4 percent water vapor.

It's Not the Heat . . .

So how does all this explain the difference between Norfolk, Virginia, and San Francisco, California? People usually think of Norfolk as more humid than San Francisco, and in terms of absolute humidity, they're right. Warm Norfolk air contains nearly twice the water vapor content of the cooler San Francisco air. But the relative humidity in San Francisco is actually higher than in

Heat Stress Index

To account for the combined effects of heat and humidity on people and animals, scientists have developed a heat stress index that gives the apparent temperatures for given conditions. For example, on a summer day in New Orleans it might be 90°F with 60 percent humidity, which together create an apparent temperature of 100°F.

Relative humidity (%)	Air temperature (°F)							
	70	75	80	85	90	95	100	105
	Apparent temperature (°F)							
0	64	69	73	78	83	87	91	95
10	65	70	75	80	85	90	95	100
20	66	72	77	82	87	93	99	105
30	67	73	78	84	90	96	104	113
40	68	74	79	86	93	101	110	123
50	69	75	81	88	96	107	120	135
60	70	76	82	90	100	114	132	149
70	70	77	85	93	106	124	144	
80	71	78	86	97	113	136		
90	71	79	88	102	122			
100	72	80	91	108				

Norfolk. The cooler San Francisco air contains less water, but the water content is closer to the maximum for that temperature. In July, the average relative humidity in San Francisco is typically around 74 percent, while Norfolk averages around 60 percent.

This belies the weather cliché, "It's not the heat, it's the humidity." It's actually both the heat *and* the humidity that can make you miserable. When it's hot, your body relies on sweating to keep cool. As sweat evaporates into the air, it draws heat from your skin. This *evaporative cooling*, as it's called, lowers your skin's temperature. Since Norfolk is hotter, you sweat more than you do in San Francisco. Although Norfolk's relative humidity is lower than San Francisco's, it's high enough that sweat doesn't evaporate quickly—it clings to your skin and you stew in your own juices.

When sweat can't readily evaporate and cool your body, the air temperature feels warmer than it actually is. Most people are comfortable when the air temperature is moderate and the relative humidity is between 40 percent and 70 percent. When relative humidity is below 40 percent, evaporative cooling from sweat can make it seem as if the air temperature is slightly cooler than it actually is. When relative humidity is over 70 percent and the temperature is 75°F or above, sweat clings to your skin, making the air feel warmer than its actual temperature.

Why are Norfolk summers so muggy and warm compared with those in San Francisco? With an abundant supply of water off both coasts, and nearly the same exposure to sunlight, you might think the temperature and humidity for the two cities would be similar. But the ocean water off Norfolk is much warmer than the ocean off San Francisco (see page 13). Flowing near Virginia's coast is the Gulf Stream, a current that brings warm water northward along the coast of the southeastern United States and helps raise summer sea surface temperatures to the low 70s near Norfolk, compared to the mid-50s near San Francisco. The prevailing summer winds also import warm, moist air from the Gulf of Mexico and the Atlantic east of Florida. The combination of a warm ocean and the inflow of warm, moist air makes the typical summer weather in Norfolk (and all along the nearby coast) hot and muggy.

On the West Coast, summertime conditions are very different. The California Current carries cool water southward along the Oregon and California coasts, cooling the air above it. The winds also tend to flow from the north, bringing cool air and causing upwelling of cool subsurface ocean water along the coast. The combination of cold water and cool air keeps the typical summer weather comfortable, though sometimes a bit chilly.

Moist air moving across the cool coastal waters contributes yet another factor to San Francisco's climate. As moist air blows over the ocean, it often cools enough for its water vapor to condense into liquid droplets and form fog. It's not unusual to see a wall of fog hugging the coastlines of California, Oregon, and Washington during much of the summer. When the sea breeze kicks in, this cool, fog-shrouded air sweeps over the land, keeping the summertime temperatures cool, and chilling the tourists in San Francisco.

Variations in Humidity

You don't have to live on the East Coast or in the South to feel the effects of warm, sticky air. If you've ever lived in or visited Cleveland in the summer, you know that the Midwest can be just as hot and muggy. Prevailing wind patterns in the summer can carry humid air more than a thousand miles from the Gulf of Mexico into the Great Plains and upper Midwest. The combination of heat and humidity in midwestern communities often sends apparent temperatures climbing into the 100°F range.

Mountain ranges can act as barriers to moisture. The desert of the Great Basin, which spans Nevada, eastern California and Oregon, and parts of Idaho and Utah, is sandwiched between the Sierra Nevada range to the west and the Rocky Mountains to the east. Most of the moist air flowing toward the Great Basin must first pass over these mountains. As the air passes over the mountains, much of its moisture falls as rain or snow, so that the air that finally reaches the Great Basin is relatively dry.

During much of the year, weather in the Great Basin is dominated by a high pressure system near the earth's surface, into which dry, high-altitude air descends. As the air sinks toward the

I n a typical summer, 100 to 200 people in the United States die from the effects of summer heat. In a two-week period in 1995, the combined effects of record high temperatures and humidity drove the apparent temperature in Chicago up to a simmering 120°F and killed nearly 500 people and 4,000 head of livestock in surrounding areas.

The southwestern deserts of Arizona, Nevada, and California typically have the hottest and most dangerous summer weather. Evaporative cooling from sweat does reduce a 110°F temperature down to an apparent temperature of 106°F, but that still makes the summer weather outside a Las Vegas casino more dangerous than a heat wave in Atlanta or Houston, where apparent temperatures rarely climb above 100°F.

Sling Psychrometer

Want to know how humid it is? Well, you could watch a weather report. . . or you could build your own sling psychrometer.

A sling psychrometer is a simple instrument that lets you determine relative humidity and dew point. To make one, first firmly tape two thermometers (that display Centigrade measurements) to a board that's big enough to hold them side by side. Position the thermometers parallel to the length of the board, with the bulbs of the thermometers extending beyond the edge of the board.

Wrap a small piece of wet cloth around the bulb of one of the thermometers and hold it in place with a rubber band. Leave the bulb of the other thermometer bare. Tie about a foot of string to the other end of the board.

Hold the string and swing this device around in front of you for a few minutes, being careful not to let it crash into anything. When your arm gets sore, stop swinging and read the temperature on each of the thermometers. The thermometer wrapped in wet cloth will probably show a lower temperature than the dry thermometer. That's because water evaporated from the damp cloth, cooling the cloth and the thermometer.

The drier the air is, the more water will evaporate and the greater the temperature difference will be. If the two temperatures are the same, then the relative humidity is 100 percent.

Use this temperature difference and the table at right to determine relative humidity. To get an accurate reading, swing your psychrometer for several minutes, until the thermometer wrapped in wet cloth stays at a constant temperature. That's when you know the cloth is as cold as it's going to get.

Psychrometric Table

To find relative humidity, check the temperature on the dry-bulb thermometer. Then determine the difference between the dry-bulb and wet-bulb thermometers. Where these two measurements intersect, you'll find the relative humidity.

Temperature on dry-bulb thermometer (°C)	Temperature differences between dry-bulb and wet-bulb thermometers (°C)								
	1.5	2.0	2.5	3.0	3.5	4.0	4.5	5.0	7.5
10°	82	76	71	65	60	54	49	44	19
15°	85	80	75	70	66	61	57	52	31
20°	87	82	78	74	70	66	62	58	40
25°	88	84	81	77	73	70	66	63	47
30°	89	86	82	79	76	73	70	67	52
35°	90	87	84	81	78	75	72	69	56

desert floor, it compresses and warms, which lowers the relative humidity of the already dry air. In most of the world's deserts, descending air is an important factor in keeping the relative humidity low and warming the air. (See page 28.)

High-elevation communities can be nearly as dry as the desert. Most of the water vapor in the atmosphere comes from the evaporation of surface waters. As this moist air rises in the atmosphere, it expands and cools. Water vapor begins condensing out of the air as it cools. The higher the air rises, the more water vapor condenses into liquid water or ice. At 5,338 feet above sea level, Casper, Wyoming, has a relative humidity of around 25 percent in the summer.

No matter what the temperature, when the relative humidity drops below 30 percent, your skin and lips are likely to dry out and crack as you lose water from your body through sweating and evaporation. Following a tour of the desert in southern Utah, Exploratorium editor Pat Murphy reported on the effects of low relative humidity. Over the course of her weeklong visit, her lips chapped, her nose chapped, and, during a bout with allergies and watering eyes, the skin around her eyes chapped. By the end of the week, she says she was "dipping her entire head in ChapStick," which cuts down on water loss by protecting lips and skin with a protective layer of oil and wax.

You don't have to visit a desert or the mountains to experience low relative humidity, however. In the winter, building interiors are often very dry even though the relative humidity outside is often within the comfort zone. Consider, for example, a typical January day in Detroit, Michigan. Outside, it's about 23°F, with a relative humidity of 70 percent. When that air is pulled in and heated to 70°F, the amount of water vapor doesn't change, but the relative humidity drops to 11 percent. At that level, water evaporates readily; houseplants require more frequent watering, wood furniture may crack, and people's eyes and sinuses dry out and become irritated.

Water Transport

At any given time, there are about 25 million billion pounds of water in the atmosphere, mostly in the form of water vapor. If it all condensed and fell to the earth's surface at once, it would cover the entire globe with an inch of water.

Most of the water in the atmosphere comes from the tropical oceans, where the sun's energy is concentrated and the ocean waters are warmest. Much of the water that evaporates from those oceans becomes rain over the sea or over tropical land masses, creating lush rain forests near the equator.

Some of the water vapor that evaporates in the tropics is transported by winds to other parts of the globe. During winter, the "Pineapple Express," low-level winds blowing northeastward from near Hawaii, may bring relatively warm humid air to California, Oregon, and Washington. Tom Murphree recalls one restless January night in northern California when he woke up hot and sweaty and went outdoors to cool off. Instead of the typical jolt of cold air, he felt a balmy wind that carried a faint flowery scent that reminded him of Hawaii, where he'd once lived. The next day he checked the weather maps, and sure enough they showed that a steady wind was blowing from Hawaii in a nearly straight path to the California coast. Tom is still hoping to find a chemist who's studied how far scents can travel on the wind to confirm his hunch that he really did smell a little bit of the islands that winter night.

Scientists have estimated that a typical water molecule spends about a week in the atmosphere before condensing and falling back to the surface. During that time, the atmosphere's typical winds would carry this typical molecule about 6,000 miles to the east or west, and 600 miles to the north or south. A molecule might leave the sea surface and enter the atmosphere somewhere off Portugal, for example, travel with the trade winds to Florida, then fall in a raindrop over Miami.

How Water Leaves the Air

Water has many escape routes out of the atmosphere. You've seen one of those routes if you've ever stopped on a morning stroll to admire a dew-covered spiderweb. If temperatures drop low enough, water vapor in the air condenses and forms dew on spiderwebs, grass, or other cool surfaces. When a surface is colder than the surrounding air, it can chill the air to the point where some of its water vapor condenses. That's why beads of water form on the sides of a cold soda can on a hot day.

As the temperature drops and the amount of water vapor in the air remains the same, the relative humidity increases. The temperature at which water vapor begins to condense is called the *dew point*. At the dew point, the air is saturated and relative humidity is 100 percent. For that reason, meteorologists often use the dew point as another measure of humidity.

Suppose the dew point is 65°F. If the air cools to 65°F, it will be completely saturated and its water vapor will start to condense. The air usually feels uncomfortably warm and humid when the dew point rises above 68°F. During the summer in steamy southeastern cities, such as Miami, New Orleans, or Mobile, Alabama, it's not uncommon for the dew points to be in the mid-70s, an indication that both the temperature and relative humidity are high. During

Dew Around the House

Y ou can observe the dew point in action every time you take a hot shower and fog your bathroom mirror. During the course of your shower, the hot water warms the air and water evaporates into the warm air. When this warm, moist air comes in contact with the cool surface of the mirror, the air cools to its dew point and water vapor condenses to form a foggy film on the glass.

By the way, you can use this condensation to keep from getting a chill after you turn the shower off. Stay inside the shower stall while you dry off. The water vapor condensing inside the stall releases latent heat to the air and to you, keeping you warmer than if you dried off outside the stall.

a record-breaking heat wave in Wisconsin in 1995, the dew point in LaCrosse peaked at 80°F, as hot and humid as a tropical rain forest.

The dew point can serve as a rough predictor of the overnight temperature. As long as the air is relatively still, the overnight temperatures will typically cool to the dew point and hold there. That's because water vapor releases latent heat when it condenses (see page 52), which helps keep the temperatures from dropping much lower than the dew point.

When the overnight temperature drops below 32°F (the temperature at which liquid water turns solid), frost may form. Frost is not dew that's been frozen, it's ice that forms when water vapor goes directly from a gas to a solid. Frost can create beautiful crystalline patterns on glass, leaves, and other cold surfaces. These patterns grow as more water vapor is deposited. Water vapor that condenses into dew above 32°F can freeze when the air turns colder, but it creates solid ice drops or a glaze of clear ice, rather than frost.

Just as water vapor can go directly from a gas to solid water, ice and snow can turn into a gas without melting first. Maybe you've noticed that piles of snow shrink or disappear during the day, even when the temperature is below 32°F. The snow is vaporizing directly into dry air, a process that meteorologists call *sublimation*. Sublimation can cause snow or ice crystals in the atmosphere to vaporize before they fall to the ground. This process also causes frozen foods in your freezer to dry out if they aren't in airtight containers and makes your ice cubes shrink, only to reform as frosty scum on your ice cream.

The Birth of a Cloud

Like most gases, water vapor is invisible. When moist air is cooled to its dew point, either at the earth's surface or in the atmosphere, it becomes saturated and individual molecules of water vapor come together and form very tiny droplets of liquid water. When water vapor condenses in the air, it forms clouds or fog (which is simply a cloud that condenses near the ground).

In order to condense, water molecules need a site where they can meet each other. Help comes in the form of tiny particles of dust, smoke, salt from ocean spray, pollen, plant spores, volcanic ash, or even debris from oceanic bacteria. The famous London fog became noticeably thicker during the Industrial Age as smoke from coal fires added more particulates to the air. These minute airborne particles, less than a millionth of an inch across, are collectively called *condensation nuclei*. They serve as meeting places for water vapor molecules, which glom onto the condensation nuclei and attract other water molecules to the growing droplet. Billions of water molecules must glom together to form a typical droplet in a cloud.

No matter how much water vapor there is in the air, clouds will not form if there are no particulates that can serve as condensation nuclei. Paul Doherty, a physicist at the Exploratorium, recalls a summer hike he took on Mount Katahdin, the highest peak in Maine. Paul remembers the air that day was very clear and warm, about 80°F, with a very high relative humidity. As he hiked up the mountain, it got progressively cooler with increasing elevation. After he'd hiked up about 2,000 feet, Paul noticed that drops of water were forming on every hair on his body until he was soaked from head to toe. He had walked to the elevation where the air was at its dew point, and his hairs were serving as nuclei for water vapor to condense directly on his body. He was, in a sense, becoming a human cloud. Yet all around him, the sky was clear.

Paul's experience was unusual because there's almost always something in the troposphere for water vapor to condense around. Salt from the ocean can blow hundreds of miles inland on the wind, and dust and other particles are also common in most places.

The formation, movement, and disappearance of clouds reveal the mysterious workings of the atmosphere. Suppose it's a warm summer afternoon and you notice a white, puffy cloud begin to form over a hot, sun-warmed surface—a black-topped parking lot, for example. As the cloud forms popcorn-like balls that grow from its flat bottom, the top edges start to fray and dissolve. After fifteen minutes or so, the whole cloud may be blown horizontally by the surrounding wind, shredding into pieces and slowly disappearing as it moves. Soon another cloud forms in its place.

In this hypothetical example, air warmed by the dark, dry surface of the parking lot begins to expand and rise. As it rises, it's surrounded by air at a lower pressure than the air nearer the surface. So the rising air can expand even more. As it expands, it cools off (see page 28). The rising air is also cooled by mixing with the surrounding air, which gets colder with increasing altitude. When the air cools to its dew point or lower, water vapor begins to condense around nuclei carried aloft with the rising air, and a cloud is born.

As the cloud continues to form, turbulent winds within the cloud cause its outer edges to mix with and evaporate into the cool, dry surrounding air, while new moisture rising from below condenses and helps the cloud to maintain itself or grow. This process tends to form discrete fluffy clouds called cumulus clouds, after the Latin word for "heap." The flat bottom of a cumulus cloud marks the level at which the surrounding air's temperature matches the rising air's dew point.

Clouds only form when the proper ingredients are in place. The main ingredients are moist air, condensation nuclei, and cooling. As we discussed earlier, moist air can come from many sources, but the most abundant source is the ocean. Places that have a steady wind blowing from the ocean or a lake often have cool, moist air and cloudy weather. Astoria, located on the northwest coast of Oregon, has 240 cloudy days a year, making it the cloudiest city in the country. Seattle, Washington, is not far behind with 229 cloudy days a year.

Deserts and other regions that are isolated from cool, moist air generally have clear, cloudless skies. The sunniest city in the U.S. is Yuma, Arizona, where the sun shines 90 percent of the time— 328 days per year. El Paso, Texas, Sacramento, California, and Albuquerque, New Mexico, all have more than 280 days of sunshine a year.

Quick Guide to Clouds

Clouds come in an amazing variety of shapes and sizes. Given this variety, it's understandable that cloud categorization is not simple. One major system is based on three common cloud shapes. Clouds with a curly, wispy, or feathery shape are called *cirrus* clouds. Lumpy clouds are called *cumulus*. Clouds that are flat or occur in layers are called *stratus*.

But naming clouds according to shape leaves out some useful information about where the clouds are in the sky. Almost all of earth's clouds lie within about 10 miles of the surface, in the troposphere. Knowing how high in the troposphere a cloud is can be useful in knowing what's in it, what's likely to come out of it, and what weather is likely to follow. So clouds are also grouped as high, middle, and low clouds, according to how high they are within the troposphere.

The convenience and simplicity of a system based on appearance, plus the usefulness of a system based on cloud height, has led scientists to settle on a hybrid system. The cloud names and interpretations in this system can still be a bit messy. But fortunately, most clouds fit rather well into one of four main groups based on height, with these groups containing a total of just 10 main cloud types.

1. High clouds are at about 15,000 to 60,000 feet. At these altitudes, the air is very cold and the clouds are made almost entirely of ice. The wind at these altitudes is usually very strong, and often smears out the clouds, giving them a gauzy or curly look. The Latin

louds can be categorized by their shape and by how high they are in the sky. Cirrus clouds are generally high and wispy, reflecting their icy composition. Cumulus clouds occur at many levels and are generally puffy, revealing the turbulent mixing of air going on in and around them. Stratus clouds are usually at low and middle levels and have a flattened look.

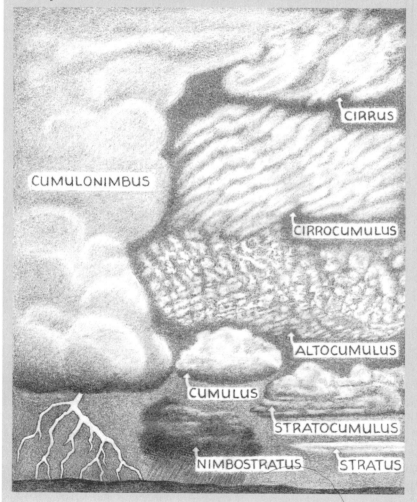

root for "curly" is cirrus, so all three of the clouds in this group start with the root *cirro*. These are thin clouds, through which the sun and moon can be seen rather clearly.

Cirrus clouds look curly, feathery, wispy, or streaky. They're very common, and their motion tells the direction of the upper tropo-spheric winds. Cirrus clouds that resemble streaming mare's tails are common, and are called "mares' tails."

Cirrocumulus are puffy clouds almost always arranged in large groups that cover a part of the sky. The puffy lumps usually have indistinct, wispy surfaces. The clear sky and pearly clouds in these groups sometimes form a diamond pattern that resembles the scaly skin of a fish, called a "mackerel sky."

Cirrostratus clouds are like a very thin sheet of gauze stretching across the sky. Sometimes, the sheet is so thin that the sky doesn't appear to have any clouds at all, but to simply be whitish. Through a cirrostratus sheet, the sun or moon may have a slightly fuzzy look and a halo due to light refraction by the clouds' ice crystals.

There is so little moisture at high levels that high-level clouds produce very little precipitation. However, when cirrus are replaced by cirrocumulus, a storm may follow, which accounts for the folk saying: "Mackerel sky, storm is nigh."

2. The middle clouds are at about 6,500 to 26,000 feet. Names for clouds in this group include the Latin root *altus,* meaning "high." These clouds are made of both ice and liquid water.

Altocumulus are puffy clouds with distinct edges usually arranged in groups. The groups often look like waves or long parallel rolls. The turbulent air motion in these clouds helps give them their lumpy appearance.

Altostratus clouds form extensive and indistinct layers that cover much or all of the sky. They're often thick enough to obscure the sun or make it look like no more than a fuzzy area that's lighter than the rest of the sky.

Snow or rain may come from both of these middle-level cloud types, but usually they're more an indicator of precipitation yet to come. Altocumulus in the morning may foretell nimbostratus and thunderstorms in the afternoon. Altostratus often appear several hours ahead of the precipitating clouds in a warm front.

3. Low clouds lie below about 6,500 feet and can occur just above the earth's surface. Because they are at a low altitude, these are relatively warm clouds made mainly of liquid water, although in winter they may also contain ice. They tend to have a layered form, usually due to overlying air that's less dense and limits their upward growth. Their names are based on the Latin root *stratus*, which means "spread" or stretched out.

Stratus clouds often cover the sky with a monotonous gray, like a smooth, thick, gray blanket. Sometimes stratus is called "high fog," but when fog clears from the surface, stratus clouds may remain overhead. As with fog, stratus clouds are common over cool surfaces.

Stratocumulus clouds are lumpy, blob-like clouds arranged in broad sheets. Sometimes they form from the breakup of a stratus layer or of a large cumulus cloud.

Nimbostratus are like stratus but darker and with rougher bottoms. They form in stable air that has been forced to rise along a front (see page 81) or where surface winds have converged.

The nimbostratus are the main rain- and snow-makers of the low-level cloud group, creating light to moderate precipitation that may last for many hours.

4. The fourth cloud group contains clouds whose distinctive feature is their vertical growth and decay. Therefore, they're called "clouds with vertical development." These clouds grow within columns of turbulent, unstable, rising air, which gives them their billowy upper surfaces, and their rapid and dramatic changes in shape.

Cumulus clouds are the clouds of many children's drawings. They look like cotton balls, often with flat bottoms and puffy tops. They often form as air rises over warm surfaces and reaches the level at which its water vapor condenses. The flat bottom of the cloud marks the height at which the air temperature matches the dew point and water vapor condenses to form a cloud.

Cumulonimbus clouds are the result of cumulus clouds that have been able to grow quite tall, up to 40,000 feet or more. They resemble great towers with dark bases, and often with flat, icy, white tops that fan out downwind of the tower.

Cumulus clouds tend to produce light, brief showers. But cumulonimbus can create extremely heavy rain and snow, along with hail, lightning, and thunder. (*Nimbus* is Latin for "rain cloud.") Cumulus clouds (including cumulonimbus) produce some of the world's most dangerous weather, from thunderstorms and tornadoes to hurricanes.

Up, Up, and Away

Clouds form when moist air cools. Most commonly, when the air is lifted away from the earth's surface, it expands and cools as it rises. This commonly happens in four different ways.

As we described on page 64, sunlight can warm an area of land or water, which warms the air above it, providing the energy needed to push moist air aloft. If the surface warming is intense enough, as it often is in the tropics and during summer in many other parts of the world, cumulus clouds can grow high into the troposphere, becoming great roiling thunderheads or cumulonimbus clouds. As the water vapor condenses in the rising air, it releases latent heat, which further warms the air and pushes the developing cloud higher into the atmosphere.

Moist air can also be lifted up into the troposphere when winds run into mountains and are deflected upward. If you've visited or lived near mountains, you've probably noticed that clouds often form in the afternoon above the windward side and around the mountaintops. It's often said that mountains create their own weather, and it's true, in large part, because of moisture that flows up the slopes and condenses into clouds. If the air is moist enough, these clouds can produce heavy rain or snow.

You can see how mountains affect rainfall by comparing precipitation in the mountains with that of the nearby flatlands. For example, the mountains just east of Salt Lake City receive much more rain and snow throughout the year than the city itself.

The western slopes of the Sierra Nevada, Cascade, and Rocky Mountains also collect clouds, rain, and snow from the winds that blow off the Pacific Ocean. The temperate rain forest on the Olympic Peninsula in Washington owes its existence to moist Pacific air rising up the windward slopes of the Olympic Mountains. It is one of the cloudiest and rainiest places in the U.S. In nearby Quillayute, Washington, it rains 210 days a year, with annual rainfall topping 100 inches.

Winds that flow up mountains generally lose their moisture before they reach the top. As they rush down the back side of the mountain, the air compresses and warms, creating the hot, dry downslope or Chinook winds. The rainiest place in the U.S. is Mount Waialeale, Hawaii, with an annual rainfall of 460 inches. This volcanic mountain lies in the path of tropical trade winds. As warm, moist air travels up the volcano, water vapor condenses to dump great torrents of rain on its windward slope 350 days a year.

The leeward side of the volcano is in what's called a "rain shadow"; the mountain shadows it from rain. There the climate is much drier than it is on the windward side. Death Valley, located in the rain shadow of the Sierra Nevada range, is the driest spot in the U.S. with annual precipitation of just 1.2 inches per year.

Mary Miller's sister, who lives in Seattle, escapes the clouds and rain on the windward side of the Cascades by driving up and over the mountains to the Yakima Valley where her family has a vacation home. The rain shadow of the Cascades creates a high desert in central Washington that's the perfect getaway spot for waterlogged coastal dwellers: the summer days are warm, clear, and dry and the nights are pleasantly cool.

The lifting action that produces clouds also occurs in places where winds flowing near the earth's surface run into each other, leaving the air with no place to go but up. The largest area of clashing winds is the Intertropical Convergence Zone, a region where converging trade winds and intense surface heating drive moist air high into the troposphere. This broad band of converging, rising air in the tropics creates the world's most active zone of thunderstorms. Another place where converging winds help create clouds and rain is over the Florida peninsula in the summer, where sea breezes from the west and east coasts of the state flow inland and meet in the middle. The clouds produced by these rising masses of warm, moist air give Florida more midafternoon rainstorms than any other place in the nation.

Finally, an entire layer of air can rise up and over another layer of air that slides underneath it. For instance, a layer of warm air flowing into a low pressure region can be wedged up by a layer of cooler air near the ground. When this happens, water vapor in the lifted layer of air may condense to form stratus clouds that can cover the sky over large areas. These flat, layered clouds generally produce light, steady rain, if they produce any at all.

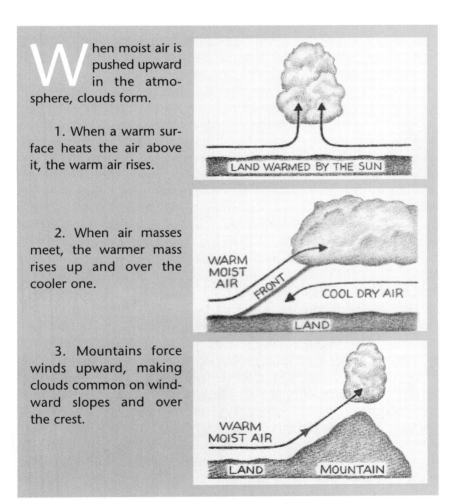

When moist air is pushed upward in the atmosphere, clouds form.

1. When a warm surface heats the air above it, the warm air rises.

LAND WARMED BY THE SUN

2. When air masses meet, the warmer mass rises up and over the cooler one.

WARM MOIST AIR

FRONT

COOL DRY AIR

LAND

3. Mountains force winds upward, making clouds common on windward slopes and over the crest.

WARM MOIST AIR

LAND   MOUNTAIN

What Goes Up Eventually Comes Down

As any weather-watcher knows, clouds do not always mean it will rain. The ice crystals or droplets of water in a cloud are microscopic, with a typical droplet measuring only about one thousandth of an inch across. Because they're so small, the crystals and droplets fall very slowly, which is part of the reason that clouds remain suspended in the air. It would take about two days for a cloud droplet to fall from the sky. Long before the droplet reached the ground, it would evaporate. To form a raindrop large enough to fall to earth, about a million microscopic cloud droplets have to somehow come together. That takes some major matchmaking.

In cold clouds, the matchmaker is a microscopic particle, called a *freezing nucleus*, that provides a surface on which water vapor can collect and freeze. This creates a tiny ice crystal, which can grow as it attracts more water vapor and droplets. Individual crystals may glom together to form snowflakes. Larger crystals and flakes will fall, and if they fall through warm enough air, they'll melt and hit the ground as rain. This can happen in any season if the clouds are cold enough. Summer rain falling from tall cumulonimbus clouds may well have started off as ice crystals and snowflakes in the icy upper regions of the clouds. This is called the *Bergeron process,* named after the Swedish meteorologist who came up with the idea in the 1920s after watching stratus clouds form during the winter in the hills near Oslo, Norway.

The Bergeron process explains how rain or snow falls from clouds that are below freezing, but what about warm clouds, like the ones that produce heavy rainfall in the tropics? A second method, called the *collision-coalescence process,* helps explain this precipitation. In any cloud, there are big droplets and small droplets. The bigger droplets are thought to originate around large condensation nuclei or water-attracting particles, such as bits of sea salt. As the bigger cloud droplets begin to fall, they can sweep up smaller droplets in their path. If the cloud is very tall, or strong updrafts keep the drops aloft, these big droplets have a better chance of colliding and merging with many small droplets and growing large enough to fall as rain.

Cloud seeding is one way to supply condensation nuclei for water droplets and ice crystals. Planes flying through clouds release particles that could serve as condensation or freezing nuclei. Some common cloud-seeding materials are dry ice and silver iodide. Despite millions of dollars spent on research into cloud seeding, there's little clear evidence that it really produces more rain and snow than nature would produce on its own.

Whether a cloud will produce snow, rain, sleet, or hail depends primarily on the air temperature. Precipitation can start as snow, but if it falls through warm air it melts into rain. Raindrops can freeze if they fall through a layer of cold air near the ground. Sleet is rain that freezes (or almost does) as it falls through a cold air layer. Freezing rain or *glaze* is rain that remains liquid in the air, but freezes on contact with the ground or other cold surface.

Hail—lumps or balls of ice that fall from the sky—are a by-product of thunderstorms. Tiny lumps of ice form in the cold tops of cumulonimbus clouds and begin falling through the tall cloud. The little lumps are slowed down or kept aloft by the strong updrafts formed during a thunderstorm. Drops of water freeze on contact with the ice, forming layers like the skin of an onion, and can grow as large as a softball before falling to the ground. Large hailstones cause hundreds of millions of dollars of crop and property damage each year. In 1986, a hailstorm over Bangladesh produced two-pound stones that killed 92 people. In 1994, a late spring hailstorm in Gillette, Wyoming, took about 15 minutes to destroy the roofs of hundreds of homes.

Hail is most common in the midlatitudes of both hemispheres, often downwind of large mountain ranges. The Great Plains of the U.S. and Canada, central Europe eastward to the Ukraine, the Himalayan region, southern China, and portions of Argentina, South Africa, and southeastern Australia are all subject to hailstorms. In the U.S., hail is most frequent in "Hail Alley," a 700-square-mile swath along the borders of Nebraska, Colorado, and Wyoming, where there are typically nine to ten days of hail a year.

Precipitation Variations

Just as clouds and precipitation can vary from place to place, they can also vary by season. Clouds and rain may be the product of local or small-scale conditions, such as local winds that push moist air up the sides of mountains or surface heating that lifts air into the troposphere to form clouds. The weather patterns produced by these local conditions change daily; an afternoon storm usually clears up by the next morning.

Sometimes large-scale processes can override these local conditions. In the summer, hurricanes that form off West Africa can push into the southeastern U.S., or cool Canadian air can flow into the eastern U.S. to break a hot, muggy spell.

In the U.S., from the Southwest and Great Plains eastward to the Atlantic, summer tends to be the season of afternoon and evening rainstorms, as warm, moist air sweeps in from the Gulf of Mexico and dumps its moisture. Along most of the West Coast, summer is the driest season. Local winds may bring in cool fog and low clouds from the ocean, but these low-level clouds produce little or no rain. However, if these winds blow far enough inland, they may produce afternoon and evening rain showers over the Cascades and the Sierra Nevada range.

Winter weather is often dominated by large-scale processes. Winter storms, which may have their origins thousands of miles from where they drop their rain or snow, are a major source of water for most of the United States. Blue Canyon, California, near the Sierra Nevada ski resort community of Lake Tahoe, lies in the path of many winter storms and is the snowiest spot in the coun-

So-called "bad hair days" often coincide with high humidity. That's because hair eagerly absorbs water, which pushes apart the molecules that make up the hair, turning curly hair frizzy and straight hair limp. In fact, hair reacts so reliably to water vapor in the air that strands of human hair are used to measure humidity in a device called a *hair hygrometer*.

skimos are said to have many different names for snow, but the Scots are pretty good at naming precipitation, too. Among our favorites are mizzle (midway between drizzle and mist), spitting rain (intermittent bursts), and smirr (light precipitation, just enough to wet things down but not enough to put up the umbrella).

try, with annual snowfall of 241 inches. As winter storms pass over lakes, they can pick up extra moisture, supplying what's known as *lake-effect snow* on the lakes' leeward shores. The two snowiest cities in the country, Buffalo and Rochester, New York, are positioned on the banks of Lakes Erie and Ontario, where they receive the full brunt of lake-effect snow.

Water on the Move

Whether you live in the snowiest city in the country or in the rain shadow of a mountain range, it's important to consider how water moves into and out of the air if you want to understand your local weather. Watching rain clouds form in the afternoon or feeling the effects of humidity on your body when you exercise gives you insight into how the atmosphere is behaving.

You can learn a lot by paying attention to humidity, clouds, and precipitation, but these aren't the only factors controlling the weather. In this chapter, we've talked about how water moves into the atmosphere, travels from one area to another, and leaves the atmosphere. These movements are helped along by differences in temperature and pressure, described in chapter 1, and the winds produced by these differences, described in chapter 2.

Wind and water, differences in temperature and pressure—all these ingredients combine to create the weather that dumps winter snow in Buffalo and summer rain in the Great Plains. In the next chapter, we'll explore how winds, water, and an unstable atmosphere combine to produce extreme weather like thunderstorms, hurricanes, and blizzards.

4 Storm
Our unstable atmosphere

THE "BLIZZARD OF '96" started like any run-of-the-mill winter
storm in the eastern U.S. On January 6, 1996, warm, moist
air from over the Gulf of Mexico and the Atlantic crept into
the deep South and traveled with a meandering jet stream up the
East Coast. At first, the storm was relatively weak. But when it met
with extremely cold, dense air from the north, the moist marine air
was lifted high into the troposphere. When moist air is lifted
upward, clouds and precipitation are likely. When the lifting is fast
and forceful, as it was in January 1996, the resulting weather is

severe and sometimes dangerous. In this case, the result was one of the worst blizzards of the century.

Blizzards are storms with wind speeds greater than 35 miles per hour, accompanied by heavy snow and extremely cold temperatures. The Blizzard of '96 dumped up to 50 inches of snow from Tennessee to Maine, hitting especially hard over the Mid-Atlantic and New England states, and breaking records in many places for single-storm snowfall. There was so much snow on the roads that the plows ran out of places to pile it up. Most of the major airports closed, along with businesses, schools, and government offices. Strong winds and heavy snow or freezing rain downed power lines all over the East. By the time the storm was over, 187 people had lost their lives.

What turned a relatively mild storm into a winter killer? The extra power of this storm came from the clash between extremes of warm tropical air and cold polar air. As a prime battleground for clashes between frigid air from the north and warm, tropical air from the south, the continental U.S. has some of world's most disastrous weather. In a typical year, we can expect 10,000 severe thunderstorms, 1,000 tornadoes, 1,000 flash floods, and 10 hurricanes.

Recipe for Disaster

Tom Murphree describes a storm as a violent and short-lived readjustment of the atmosphere toward a more stable condition. When cold air and warm air meet and start mixing it up, the natural tendency is for the relatively buoyant warm air to come to rest on top of the heavier cool air. That's the stable condition. Blizzards, hurricanes, thunderstorms, and tornadoes are all part of the process of getting to that stable condition.

To get a really good storm going, you need a special set of conditions. First you need an influx of relatively warm, humid air flowing near the ground. Because warm air is buoyant and tends to rise above colder, denser air, warm air near the ground is unstable. Like a house of cards that will tumble when bumped, warm air tends to rise rapidly with just a little nudge.

As moist air rises, it expands and cools, and water vapor condenses. When water condenses, it releases what is called "latent heat." (See page 52.) This latent heat warms the air further, which feeds the growing storm by allowing moist air and clouds to rise higher into the troposphere before releasing their load of rain, snow, hail, or other precipitation. When you see clouds that keep rising higher in the sky, you're seeing a result of an unstable atmosphere.

Like a house of cards that will tumble when bumped, warm air tends to rise rapidly with just a little nudge.

To keep a storm brewing, you also need conditions in the upper air that allow the rising air to spread out and make room for more warm air to rise. Changes or disturbances in the high-atmosphere winds may block the upward flow and keep a lid on the storm, or they can help move rising air from a developing storm out of the way. That's why a TV weathercaster may describe an upper-air disturbance as a sign of a developing storm.

To grow and maintain themselves, storms need energy. Ultimately, most of the energy for storms comes from the warmth of the sun. Solar warming leads to evaporation of water. Latent heat released by condensing water vapor is a primary energy source for violent storms, including blizzards, thunderstorms, and hurricanes.

A storm's energy can also come from the pressure and temperature differences between air masses. An air mass is a huge blob of air, usually about a thousand miles across and several miles thick, that has distinctive conditions of temperature, moisture, and pressure. The more extreme the contrast between the two meeting air masses, the more forcefully the warm air mass will be driven upward, and the more violent the storm will be. In the central and eastern U.S., some of the most vicious thunderstorms and tornadoes occur in the spring when temperature contrasts between still-wintry northern air and rapidly warming southern air are the greatest.

Air Masses on the Move

Changing weather, especially extreme weather, can often be traced back to invading air masses from distant locales. Air masses are large enough that they retain the character of the place where they formed, at least for a while. An air mass that forms over a tropical ocean, for example, is warm and moist; one that forms over a snow-covered northern continent is cold and dry.

If you know where an air mass is from, you know something about how it may affect your local weather. Unseasonable weather is often the result of an air mass passing through. For instance, a continental polar air mass formed over northern Canada can flow southward over the Great Plains and Midwest. In winter, this results in freezing-cold winds blowing across the eastern U.S. In summer, this same movement can bring the welcome relief of cool, dry weather to the upper Midwest, at times extending all the way down into Texas and Florida.

Maritime polar air masses, formed over the oceans, are warmer and wetter than continental polar air masses. Air masses that form over the northern Pacific often bring damp, chilly weather into the Pacific Northwest and heavy rain and snow to the mountains downwind.

Because winds and weather generally flow from west to east across the lower 48 states, Pacific air masses influence weather to a greater extent than air masses from the Atlantic. Occasionally, winter winds flowing out of a North Atlantic air mass can blow into southeastern Canada and coastal New England, bringing on a dreaded Nor'Easter storm, with heavy precipitation and strong northeast winds. In the summer, polar maritime air from the North Atlantic can bring cool, clear weather to the Northeast.

The maritime tropical air masses formed over the Gulf of Mexico and the Gulf Stream are among the most influential weather-makers in the eastern two-thirds of the U.S. These air masses provide much of the moisture that makes this part of the country so much wetter than the West.

Air masses are not immune to changes in their temperature or humidity. Continental polar air will gradually warm as it moves south in the winter. By the time it has traveled thousands of miles over increasingly warm ground, an air mass that started at 10°F in the Yukon may climb to 50°F by the time it arrives in Tennessee. But if that same air mass traveled mainly over snowy ground, it could retain much of its arctic nature and freeze the tourists in Nashville.

Battlefronts and Air Wars

Air masses move for the same basic reasons any blob of air moves: They're pushed by pressure differences, with Coriolis effects helping direct the motion. Global pressure and wind patterns drive polar air masses toward the equator and tropical air masses toward the Poles, with the midlatitudes smack in the middle. When two air masses meet, the boundary between them, called a *front,* is where the most violent action happens.

When the leading edge of a warm air mass, called a *warm front,* moves into an area, it floats like a balloon, riding up over cooler, denser air already there. Wispy cirrus clouds high in the sky can be the first sign of an approaching warm front. Jet contrails that linger in the air are another indication that relatively warm, moist air has moved into the upper troposphere. Jet exhaust provides condensation nuclei that allow water vapor to condense into clouds, which will linger only when the air is humid.

As a warm air mass approaches, warm, moist air being lifted along the front often creates extensive clouds and, possibly, steady light precipitation ahead of where the front intersects the ground. As the front advances, the clouds become thicker and lower, with heavier rain, snow, or other precipitation.

After a warm front has passed, the temperature usually increases as the warmer air mass extends down to the earth's surface and replaces cooler air. If the warm air mass originated over a desert or other dry area, a warming trend may be one of the only signs that a front has moved through, since dry air won't produce clouds and rain even when it's lifted high in the troposphere.

Cold fronts usually bring more dramatic changes to the weather than warm fronts do. Cold, dense air acts like a shovel, pushing

Warm and Cold Fronts

Where two air masses of differing temperature meet, you get what meteorologists call a front.

Warm front

If a warm air mass runs into a cold air mass, the warm air rides up and over the colder air, often creating extensive clouds.

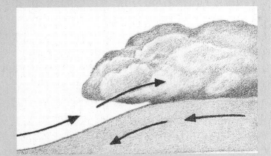

Cold front

If a cold air mass runs into a warm air mass, the cold, dense air shoves the warm air high into the troposphere, often bringing dramatic storms.

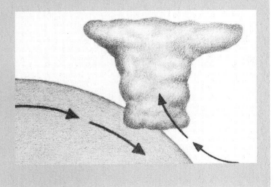

under warm air and shoving it high into the troposphere and leading to rapid cooling, condensation, and heavy precipitation. Forceful lifting, together with an unstable atmosphere, can mean the difference between a steady light rain and a gully-washing thunderstorm.

If you see "mare's tails," long wisps of streaky cirrus clouds that have a distinctive hook at one end, you may be seeing an early sign of an incoming cold front. Another indication is a "mackerel sky," in which there are closely spaced clouds named for their resemblance to fish scales. Generally, the clouds that form along cold fronts are higher than they are wide, and precipitation comes more suddenly than it does from a warm front.

Knowing how cold and warm fronts move and interact helps explain why the Blizzard of '96 was so devastating. The warm, moist air that moved up from the Gulf of Mexico and the Atlantic on January 6 brought gentle rain when it slid on top of cooler air over the southeastern states. But when the relatively warm air moved farther north and encountered a mass of freezing arctic air that had settled over the East, the mild, humid air was quickly pushed high into the troposphere, fueling massive snowstorms. A few weeks later, a warm air mass from the south spread into the same areas affected by the blizzard. This warm air and a mild storm system brought rain and record snowmelt to the area, overwhelming rivers and causing massive flooding in Pennsylvania, Ohio, West Virginia, Virginia, Maryland, New York, Vermont, and the District of Columbia.

Battle Lines and Storm Fronts

If you've ever looked at a weather map of the U.S. or other mid-latitude area, you've probably noticed that fronts usually don't advance as straight lines. When warm and cool air masses meet, they don't immediately rearrange themselves with the cool air on the bottom and the warm air on top. Instead, as they approach each other, Coriolis effects give the conflicting air masses a spin— a counterclockwise spin in the Northern Hemisphere and the opposite in the Southern Hemisphere. (See pages 40–42 and 47.) These swirling air masses form a low pressure system that can be more than several hundred miles wide. By forcing the air masses to spiral around each other rather than quickly resolving their differences, Coriolis effects prolong the battle and help large, intense storms to develop. When you see one of these low pressure systems in a satellite image, the fronts show up as cloudy curves, extending out from the center where the pressure is the lowest.

The spiraling air masses produce what's known as a *midlatitude cyclone*. (Tropical cyclones—commonly called typhoons or hurricanes—also have winds that spiral into a central low pressure center, but these storms don't involve such distinct air masses and fronts.) Midlatitude cyclones generally move eastward, guided by a jet stream. When one of these storm systems passes over your area, you might first experience a warm front, followed by a cold front, as the whole system rotates and moves eastward. Precipitation falls along both fronts. The maps on pages 84–85 trace the path of a typical midlatitude cyclone.

Winter storms that sweep across the U.S. often begin when a cold, dry air mass flowing out of Siberia or China clashes with warmer moister air over the warm ocean current east of Japan, called the *Kuroshio.* Jet streams steer many of the resulting storm systems across the North Pacific and into Oregon and Washington. As a result, cities like Portland, Oregon, and Tacoma, Washington, get most of their rainfall from November through March. As these storms rise up over the Cascade and Rocky Mountains, they dump more rain and snow.

Continuing east over the Rockies, the weakened Pacific storms gain strength over the Great Plains as cold, dry air from Canada clashes with warm, humid air from the Gulf of Mexico. This can sometimes turn otherwise mild snow flurries into major snowstorms over the central U.S.

When the Jet Gets Kinky

The storm tracks of midlatitude cyclones follow the upper-level winds of the jet streams. As a jet stream moves from the West Coast to the East Coast, its movement often resembles an out-of-control fire hose that veers first to the north, then to the south, responding to variations in atmospheric pressure in the upper atmosphere. This jet stream roughly marks the boundary between cold polar air masses and warm tropical air masses. As these air masses mix, the jet develops kinks or bends. Each bend, or change of direction in the jet, affects weather on the earth's surface, which is why TV meteo-

The Path of a Storm

These weather maps follow the course of a strong spring storm as it crosses the United States. Before it hit the Midwest, this storm system brought rain to Portland and snow to the Cascade range and beyond to the Rockies. The storm weakened in the mountains as it dumped its load of moisture, but regained strength in the Great Plains as it encountered a warm air mass from the Gulf of Mexico and a cold air mass from Canada.

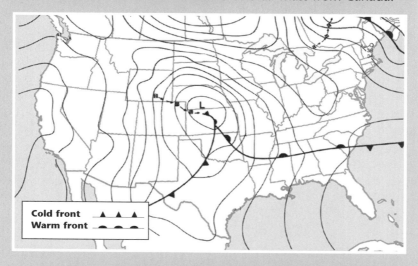

Cold front ▲▲▲
Warm front ●●●

Day 1: March 23, 1975

The first map shows a well-developed storm with a low pressure area over Nebraska and Kansas. The warm front was crossing Wichita, Kansas. Along the cold front, a line of fierce thunderstorms, called a *squall line*, had developed across Texas and Oklahoma. In Fort Worth, Texas, rolls of dark clouds signaled that the squall line was approaching. In between the cold and warm fronts, New Orleans, Louisiana, was enjoying a respite of relatively warm, clear weather.

Day 2: March 24

The powerful storm moved east, spinning counter-clockwise around a low near the border of Iowa and Minnesota, spawning blizzard conditions near Duluth, Minnesota. Meanwhile, the cold front surged through Tennessee and Alabama. After passing through Texas, the weather in that state was clear and cool.

Day 3: March 25

Traveling northeast, the central low pressure and warm front moved through the Great Lakes and New England, heading out to sea. To the south, the cold front passed through northeastern Florida, generating a small tornado and some minor hail damage. Meanwhile, a new storm was moving in from the Pacific.

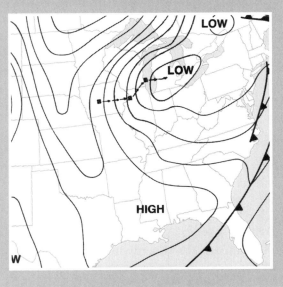

STORM

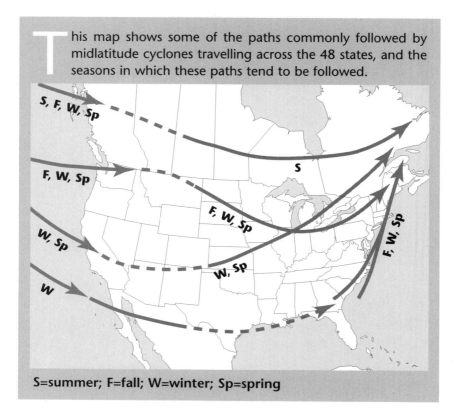

This map shows some of the paths commonly followed by midlatitude cyclones travelling across the 48 states, and the seasons in which these paths tend to be followed.

S, F, W, Sp

F, W, Sp

W, Sp

W

S

F, W, Sp

F, W, Sp

W, Sp

S=summer; F=fall; W=winter; Sp=spring

rologists will often start their forecasts with a discussion of what the jet streams are doing.

Where a Northern Hemisphere jet dips to the south, it forms what's called a *trough*. Where it bends to the north, it forms a *ridge*.

Troughs and ridges affect whether air rises up from below or sinks toward the surface. Air tends to rise and low pressure areas tend to form in the area beneath the downwind side of the trough. This low pressure contributes to the development of midlatitude cyclones.

Air tends to descend and high pressure areas tend to form in the area beneath the downwind side of a ridge. This contributes to the formation of a high pressure system out of which winds blow, called an *anticyclone*. So the bends in jet streams play an important

WATCHING   WEATHER

role in the development of surface pressures, winds, and storms.

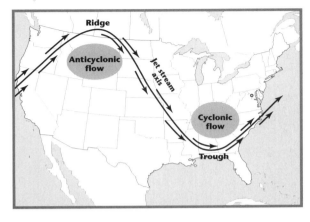

Anticyclones and cyclones come in pairs: The anticyclones feed air into the cyclones and the cyclones provide air from the anticyclones with a place to go. A jet stream supports these spinning partners as it curves above them on its wavy path. Jet streams outside the tropics are usually stronger, kinkier, and more unstable in the winter, so they have more influence over winter weather than summer weather.

Ridges and troughs usually drift east across the country, along with their accompanying storms. Sometimes, however, the troughs and ridges stay in place, forming a stationary pattern. When that happens, the normal movement of storms is altered, with some areas getting more storms than usual and others less. This is known as a *blocking pattern* because normal storm patterns are disrupted. A blocking pattern can lead to weather extremes: droughts and hot weather in one place, heavy rain and flooding in another, unusually cold weather in still another.

In the spring and summer of 1988, for example, a persistent ridge settled over the northern Great Plains. The ridge diverted moisture-bearing storms into Canada and helped block the inflow of moist air from the Gulf of Mexico. This led to the driest summer the Midwest had experienced since 1895; the resulting drought devastated corn, soybean, and wheat crops in the grain belt, and dropped the level of the Mississippi River so low that it was unnavigable.

The summer of 1993 saw a different pattern with opposite results when an upper-air trough developed over the West Coast. This giant area of low pressure helped route storms and moist air directly into the Midwest for several months. Torrential rainfall in

Lightning

Terrifying as lightning can be, it's also very common. In fact, about 200 lightning bolts will strike the earth in the time it takes you to read this sentence.

A *lightning bolt* is an electrical current—a flow of electrical charges—that travels along a path of air molecules that have been ionized or ripped apart. Most lightning develops during unstable weather conditions, when warm, moist air rises from the earth's surface to form the towering cumulonimbus clouds of thunderstorms.

Scientists are still arguing about exactly what happens in these clouds, but somehow negatively charged electrons are pulled from their atoms. The top of the cloud becomes positively charged, while the bottom of the cloud is mostly negatively charged with areas of positive charge. Eventually, the difference in charge becomes so extreme that negative charges rush toward positive charges, forming a bolt of lightning—either within the cloud or between the cloud and the ground.

Lightning simultaneously creates a brilliant flash and an acoustic shock wave, better known as *thunder*. Since sound travels about a million times slower than light, there is usually a delay between when you see the lightning flash and when you hear the thunder. If you count the number of seconds between the flash and the thunder and divide by five, that's how many miles you are from the strike. If you're more than 20 miles away, you may not hear any thunder at all. (When lightning is so far away you can't hear its thunder, it's called *heat lightning*.)

If you get caught outside during a thunderstorm, stay low. Lightning generally strikes the tallest thing around. For the same reason, don't take shelter under a tree. Cars and buildings equipped with lightning rods are safer places to be during an electrical storm. If you're indoors, remember that lightning can travel through plumbing and electrical or telephone wires—stay off the telephone and save your washing up for later.

the Mississippi River basin during the spring and summer of 1993 caused many rivers to overflow their banks. The Mississippi swelled to seven miles across in some areas. Record rains in the Midwest produced such extensive flooding that much of Iowa became a giant shallow lake. In all, the area suffered $20 billion in property damage and crop losses.

Scientists are still working out the causes for blocking patterns like these. In many cases, they've traced the origin of these unusual weather patterns to changing sea surface temperatures in the tropical Pacific. These ocean changes, known as El Niño and La Niña events, will be covered in chapter 5.

Rapids and Whirlpools in the Jet

Air flowing around a kink in a jet stream can sometimes get pinched off to spin around on its own, like a whirlpool in a moving stream. An isolated ridge, known as a *cut-off high,* can help keep clouds and precipitation at bay. An isolated trough, called a *cut-off low,* can also hang around for days, helping to maintain stormy weather.

Sometimes a jet loses its kinky nature and speeds west to east in what's called *zonal flow.* When that happens, storms from the North Pacific are likely to travel rapidly across the U.S., dumping rain or snow in the storm path and leaving areas to the north or south dry. For example, a persistently zonal subtropical jet and storm track during the 1991–92 winter resulted in a very wet winter over the southern states and dry conditions to the north.

Much of what characterizes the weather and climate of an area can be tied to semipermanent cyclones and anticyclones created by surface heating, upper-level winds, and other factors. In the summer, a large subtropical high pressure system, called the Bermuda or Azores High, parks east of the U.S. over the North Atlantic. As winds sweep clockwise around the high, they carry warm, moist air from the south up along the Atlantic seaboard. This steady supply of tropical maritime air feeds summer thunderstorms throughout the Southeast.

Pennies from Heaven

iologist A. D. Bajkob was having breakfast with his wife in a restaurant in Marksville, Louisiana, when the waitress informed him that fish were falling from the sky. He went outside and discovered fish—"from two to nine inches in length"—in the street, in the trees, and on the roofs of houses. Bajkob collected some of the fish and noted that they were "freshwater fish native to local waters . . . large-mouth bass, goggle-eye, and hickory shad."

There are actually dozens of reports of this kind. In 1924, small fish were found scattered on New York City streets after a heavy summer shower. Residents of Jelalpur, India, watched fish fall from the sky midday in February of 1830. In 1931, *The New York Times* reported that a rain of perch in Bordeaux was so heavy that it stopped traffic. Although reports of small fish are most common, other reported rains have included frogs, toads, and hazelnuts.

One proposed explanation for these fishy downpours involves waterspouts and tornadoes that touch down over water. The winds and low pressures of a waterspout or a tornado could carry water and schools of fish swimming near the surface of an ocean or lake up into a storm cloud. Once lifted, 100-mph updrafts—the kind that keep golf ball–sized hail in the air—could keep the airborne fish aloft, carrying them inland to fall with the rain when the updrafts subside.

There's a similar anticyclone, called the North Pacific High, off the West Coast. Hawaii, which is at about the same latitude as Florida, sits under the southern side of this high during the summer. The sinking air in the high pressure region helps keep air from rising high into the troposphere. This makes it difficult for Hawaii to brew the kind of strong thunderstorms seen in Florida. Hawaiian rains are gentle by comparison, with much less thunder and lightning.

Thunderstorms

You can see the path of the jet streams and advancing cold fronts on a newspaper weather map or a TV weather report. But the view from the ground provides a very different picture of storm development.

Take, for example, the development of a thunderstorm in Ohio. On a sultry summer afternoon in Cincinnati, you might notice clouds beginning to build in the west as an advancing cold front wedges under and lifts air warmed by the ground. For the first hour or so, nothing falls from the clouds; they just keep getting taller in the sky. The clouds grow massive cauliflower tops and you may see sheets of rain falling from the bottom layer and evaporating into the air.

As you look toward this approaching storm, you feel a cool wind on your face, like a breath of fresh air. This breeze is a down-draft of cool, dry air being forced down and away from the storm.

A wedge of clouds may form along the front of the advancing storm, shaped like a giant roll. You see flashes of lightning, then a few seconds later you hear the boom of thunder. As the thunder-storm approaches, the flash and boom come closer together.

When the storm passes overhead, you might experience a few minutes of pea-sized hail, then heavy rain. An hour or so later, the storm has passed to the east and the air feels much cooler and drier. The skies clear and the sun comes out again.

Most of the 100,000 thunderstorms the U.S. experiences every year are relatively mild, lasting anywhere from 20 minutes to three hours. Ten thousand thunderstorms a year are rated as severe—which means they have three-quarter-inch hail and wind gusts of 58 mph or more. These big storms can last for several hours and travel 200 miles or more.

Severe thunderstorms need both rising and descending air to stay alive. Converging winds, updrafts, and latent heat released from the condensing water vapor can push the developing clouds five to eight miles high. Fresh supplies of moist air flowing into the bottom of the storm fuel the updraft and the development of huge anvil-shaped cumulus clouds. When the clouds grow

Tornado and Hurricane Safety

B oth hurricanes and tornadoes can bring winds of over 100 mph—strong enough to threaten life and limb. If you're ever caught in such a tempest, here's what to do:

• If you're in a car or mobile home, get out and seek sturdier shelter. Don't try to outrun a tornado in your car—you're safer getting out and hunkering down in a ditch.

• In a house, the safest place to be is in the basement. If there isn't a basement, choose an interior room or hallway on the bottom floor, as far from any windows as possible.

• Don't waste time opening windows. This won't keep your house from being torn apart.

high enough to become very cold, often well below freezing, rain and hail start falling in earnest and lightning and thunder are at their peak.

The vertical circulation in a thunderstorm can be especially strong when middle-level winds flow into the storm. Water droplets in the clouds evaporate into the drier, incoming air, and this evaporation cools the air. The denser, cool air descends in rapid downdrafts at the storm's rear flank. In powerful storms, these downdrafts, also called *downbursts* or *microbursts,* can reach speeds of over 100 mph and can force airplanes from the sky. When these downbursts hit the ground, the air spreads out in a starburst pattern. Wind damage from a severe downburst can resemble the pattern left by a bomb, snapping mature trees in half and flattening buildings.

Some of the air from a downburst spreads out in front of the advancing storm, creating a *gust front,* a boundary between the cool, dry descending air and the warm, moist surface air. (The passage of the gust front provides the breath of cool air that you may feel as a storm approaches.) The cool, dry air plows under the warm, moist air on the ground, forcing it up into the thunderhead and keeping the storm fed with moisture and latent heat.

A severe thunderstorm begins to dissipate when the cooling downdrafts overwhelm the warm updrafts. As this happens, the supply of warm, moist air is cut back, depriving the storm of its primary energy source. Rain may continue for a while longer, but eventually the cumulonimbus clouds disperse and the sky clears again. Thunderstorms and other storms can also die a premature death when upper-level winds blow away the tops of the clouds and break up the storms' vertical circulation.

Twisters: Dangerous Offspring of Thunderstorms

One of the most terrifying and life-threatening products of a thunderstorm is the whirling column of air known as a tornado or a twister. An F-5 twister, like the one that ripped through the town of Jarrell, Texas, on May 27, 1997, has wind speeds of over 260 mph, strong enough to lift a house off the ground and a railcar off its tracks. In Jarrell, the winds carried a 10-ton combine a distance of 75 yards and left it in a heap of twisted metal. The greatest danger to human life in a twister is not the wind itself but the flying debris.

Fortunately, killer tornadoes are rare. Of the 1,000 twisters reported in the U.S. every year, only 2 percent are rated as F-4 or F-5, the most violent categories. The average annual U.S. death toll from tornadoes is about 80, although that number is dropping as forecasting methods and advance warnings improve.

Tornadoes can happen wherever thunderstorms occur, but they're most common in the late spring in the part of the U.S. known as Tornado Alley. This belt of tornado-prone territory starts in Texas and extends into Oklahoma, Kansas, Nebraska, and South Dakota. May is the peak month for twisters in Tornado Alley, and most of them strike during the warmest part of the day, when conditions are ripe for severe thunderstorms.

Storms and Wind Speed

One out of every 10 tropical storms develops winds of 74 mph or more, at which point the storm becomes a hurricane. Hurricanes are rated by their wind speed. Category One hurricanes are considered minimal with wind speeds between 74 and 95 mph. Category Three and above are considered major hurricanes. Category Five hurricanes are catastrophic, with winds greater than 155 mph.

Tornadoes are classified according to the Fujita Intensity Scale. They range from an F0, which has winds of less than 72 mph and causes light damage, to an F5, which has winds of more than 260 mph and causes incredible damage.

A tornado or a twister is a rotating column of air known as a *vortex*. Twisters usually form on the flanks of full-blown thunderstorms. The vortex is often created as part of a thunderstorm when swift winds above the ground and slower winds near the ground set the air between them spinning (similar to the way you can roll a pencil between your palms when you slide one hand away from you faster than the other). The vortex first spins around a horizontal axis, but when it flows into the updraft region of a thunderstorm, the wind column tips up and begins spinning around a vertical axis.

The high winds in a tornado are due to an extreme low pressure that forms at the center of the spinning column of air. Air converging toward the low speeds up faster and faster as it gets closer to the center, similar to the way a figure skater's spin increases as he tucks his arms against his body. The vortex may form a funnel cloud above the ground, which often becomes dirty from debris blown into the vortex. The funnel cloud can stay aloft or it may narrow and stretch into a faster-spinning tornado that touches the ground.

Name That Storm

The practice of naming tropical cyclones began unofficially during World War II, when military meteorologists named storms after their wives or girlfriends. The practice was adopted officially by civilian forecasters in the 1950s. Naming storms helps prevent confusion when meteorologists are tracking several storms at once.

For Atlantic storms, there are currently six different alphabetical lists that are reused in a six-year cycle. In 1979, gender equality was achieved when men's names were added to the lists.

A tropical cyclone earns a name when its winds exceed 39 mph. If a storm turns out to be extremely severe and causes unusual damage (such as Hugo in 1989 and Andrew in 1992), its name is retired from the list and replaced with a new one.

If you see a strong whirlwind of dust on the ground or a descending funnel cloud, a tornado may be approaching. Tornadoes usually travel from the southwest to the northeast, so if you're south or west of the storm you're probably in the clear. But don't depend on it; tornadoes can move erratically and at great speeds (over 30 mph).

A less destructive cousin to the tornado is the waterspout, a whirling wind that occurs over water. Waterspouts are usually much weaker and develop under milder conditions than tornadoes. Some get their start when wind flowing over the ocean blows past an island to form eddies in the air. As air is drawn into an eddy, it can rise and expand enough to condense its water vapor, making a visible funnel cloud of water droplets.

The Florida coastline is a great place to see waterspouts. Tom Murphree spent a summer working in the Florida Keys and was thrilled to see two waterspouts his first day there. By the time he left two months later, however, he was so accustomed to seeing them that he barely took notice.

Hurricanes

Tornadoes are among the smallest spiraling storms and mid-latitude cyclones are the largest. Intermediate in size between the two are *tropical cyclones.* In their mature stage, tropical cyclones are known as hurricanes in the North Atlantic and eastern Pacific, typhoons in the western Pacific, and cyclones in the Indian Ocean. Australians sometimes call them "willy-willies." These storms can stretch across a thousand miles and extend to the top of the troposphere, containing more than a million cubic miles of atmosphere.

Tropical cyclones are essentially an organized swarm of thunderstorms along spiral bands that extend into a low pressure center. These storms originate over warm tropical waters and get their energy from the warm, moist air rising from the ocean's surface. To develop, hurricanes usually need sea surface temperatures of 80°F or higher. Most tropical cyclones form when ocean temperatures peak in late summer and early fall. These storms move generally westward with the trade winds, but often turn northward and get carried eastward with the midlatitude westerlies.

To be classified as a hurricane or typhoon, the storm has to have winds that travel faster than 74 mph. In the most powerful hurricanes, wind speeds top 150 mph. Tropical cyclones can dump enormous amounts of rain and cause severe flooding if they come ashore. In 1979, Tropical Storm Claudette dropped an incredible 43 inches of rain near Alvin, Texas. Alberto, the first tropical storm of the 1994 hurricane season, dumped 21 inches of rain, causing extensive flooding in Alabama and Georgia, and well over $500 million in total damage costs.

Alberto was a small blip on the radar screen compared with Hurricane Hugo, the first hurricane in the U.S. to top $1 billion in federal emergency funds and $7 billion in total damages. Hugo started quietly enough: The storm's first stirrings were a cluster of thunderstorms detected by weather satellites on September 9, 1989, as the storms moved westward from the coast of Senegal in West Africa. Meteorologists knew to keep an eye on storms that form in this region of the tropical Atlantic, the birthplace for many of the most powerful tropical storms that eventually strike

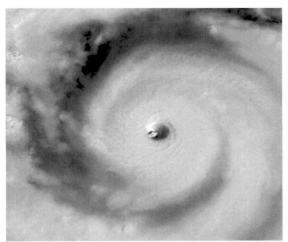

This satellite image from September 12, 1997, shows Hurricane Linda, one of the strongest hurricanes ever to develop in the eastern Pacific Ocean. The storm had wind speeds of 185 mph and gusts to 220 mph.

the Caribbean islands and the southeastern U.S. But in early September, it was too soon for anyone to know that these poorly organized thunderstorms would become one of the costliest hurricanes in U.S. history.

As the trade winds propelled these thunderstorms across the Atlantic, the storms gained strength. Storms like Hugo are fueled by moist air from the surface of the ocean. As the humid air rises inside the storm, water vapor condenses, forms clouds, and releases latent heat, warming the air further and making it more buoyant. Upper-level winds above the ocean nurtured the developing storm by spreading the rising air out of the way, allowing the air to keep rising. Surface winds flowing into the low pressure area spiraled counterclockwise as they approached what would become the eye of the hurricane. At this point, the swirling band of thunderstorms was called a *tropical depression,* the weakest version of a tropical cyclone.

Tropical cyclones develop an eye when air that has risen within the storm bands descends at the center. The descending air compresses, warms, and creates clear skies. Around the eye, a ring of cumulonimbus clouds forms a towering eyewall, leaving a 10- to 50-mile-wide column of clear, calm air at the center. The eye of a full-fledged hurricane is big enough for a plane to fly into. When the eye passes over land, it can trick people into thinking the storm is over, when half of the hurricane is still to come.

Water temperature, surface winds, and upper-level winds all cooperated to produce Hugo. When the winds topped 40 mph, it was classified a tropical storm and the National Hurricane Center in Miami gave it the name Hugo and began issuing advisories on its route and projected strength.

By September 13, 1989, Hugo had strengthened from a tropical storm to a major hurricane and was about 1,200 miles east of the Caribbean islands of Guadeloupe and Antigua. After it smashed into Guadeloupe and Puerto Rico, forecasters at the National Hurricane Center warned residents of South Carolina that Hugo had veered northeast and was heading straight for them.

Ninety percent of the fatalities from tropical cyclones are caused by coastal and inland flooding.

On September 22, Hugo hit the South Carolina coast near Charleston. With winds of over 125 mph, Hugo ripped apart houses, snapped trees in half, and flattened 9,000 square miles of forest in the Carolinas.

Bad as hurricane winds are, rising ocean waters pose an even bigger threat to property and human life. Ninety percent of the fatalities from tropical cyclones are caused by coastal and inland flooding. Low atmospheric pressure at the eye of the hurricane creates a bulge in the sea surface. If a hurricane strikes when the tide is high, wind-whipped waves, together with this bulge, create a *storm surge,* which can tower 25 feet above the ground. When it reaches shore, a storm surge can demolish buildings, erode beaches, and wash out roads. When Hugo struck the barrier islands of Charleston, its 20-foot storm surge carried beachfront houses two blocks inland.

Hurricanes usually weaken over land because their main energy source, warm ocean water, is cut off. But Hugo was exceptional. Even after it traveled nearly 200 miles overland to Charlotte, North Carolina, the winds maintained gusts of over 100 miles per hour. Some of the wind damage caused by hurricanes can come from tornadoes formed from the hurricanes' thunderstorms.

Where Hurricanes Strike

Because of their special requirements for growth and development, tropical cyclones don't threaten all coastal communities. They're conspicuously absent near the equator because Coriolis forces there are too weak to create a powerful, organized, and spinning system. The very warm waters of the tropical western Pacific breed more hurricanes (or typhoons, as they're called there) than anywhere else. Hurricanes also form in the eastern Pacific off the west coast of Central America, but few of those storms reach the U.S., since winds and ocean temperatures generally guide them westward into the open ocean. The hurricanes that reach the West Coast weaken and die over the cool waters off California, although their remnants can sometimes dump heavy rain in southern California, Arizona, and New Mexico. Prevailing winds and the cooler waters surrounding Hawaii usually insulate these islands from hurricanes. But occasionally, winds or unusually warm sea surface temperatures allow a hurricane to reach Hawaii. Hurricane Iniki, for example, struck the island of Kauai in September 1992.

Florida is one of the most vulnerable areas in the U.S. for hurricanes, attracting more than its share of deadly storms. The Florida Keys have been hit by over 40 hurricanes in the last century. In the last 30 years or so, Puerto Rico, the coastal Southeast, and the Gulf states have been a primary target of powerful hurricanes. The Northeast is much less affected because it's farther from the warm ocean waters.

The 1970s, '80s, and early '90s were a quiet period for Atlantic hurricanes. Scientists think that this respite may be coming to an end as we enter into a cycle of more frequent and more destructive hurricanes. They're still working out the global patterns that affect hurricanes, but monsoon rains in western Africa, warmer sea surface temperatures in the tropical oceans, and changing global wind patterns may all play important roles. The challenge of interpreting global patterns to forecast hurricanes and other storms will be covered in the next chapter.

Forecast

Predicting the weather

YOU STEP OUT THE DOOR and within moments you sense a storm is coming. There's just something in the air. Maybe it's the rich and earthy smell in the quickening breeze. The air feels expectantly heavy and humid to you and your joints may ache a little. If you are in the country, you may hear frogs croaking in a nearby pond and notice that a flock of geese is flying closer to the ground than it usually does. In the yard, leaves are flying by on a wind that's coming from the southwest.

These natural signs can all foreshadow an approaching storm. To forecast local weather, you need to note changes in temperature, humidity, air pressure, wind speed and direction, and sky conditions (clear or cloudy, how much of the sky is covered by clouds, what kinds of clouds you can see, how they're moving). You can go a long way toward assessing these elements using just your senses and by paying attention to the world around you. Your observations, combined with some knowledge about weather patterns, can help you predict the local weather.

When the weather is fair, birds fly higher in the sky.

Meteorologists use the same basic elements to predict the weather. Rather than restricting themselves to the local environment, however, professional weathercasters take a global perspective as well, making use of remote sensing technology, observations from weather stations all over the world, and supercomputers. With these tools, they attempt to predict the weather for the next few days and for months to come.

Natural Signs

If you don't have a supercomputer—or even a weather vane—you can anticipate changes in the weather by paying attention to changes in the natural world. Some changes indicate decreasing pressure and increasing humidity, signs that a storm may be on the way. Others indicate increasing pressure and decreasing humidity, signs of approaching fair weather.

Geese and other birds may respond to changing air density, pressure, and winds by adjusting how high they fly. When the weather is fair, the pressure is high, and birds fly higher in the sky. When the pressure is low and there are strong gusts of wind and turbulent eddies, birds stay put or fly closer to the ground.

Your joints and nerves may register falling air pressure when dissolved gases in your blood and other body fluids come out of solution to form bubbles that trigger arthritis pains and irritability.

The same drop in atmospheric pressure can cause bubbles to form in a calm body of water as gases released from decaying plants on the bottom of a pond expand and float to the top. Since decreasing pressure means unstable weather, these bubbles may indicate that rain is on its way.

The lower atmospheric pressure and higher humidity as a storm approaches can release aromatic oils from the ground, imbuing the air with an earthy smell. Water vapor in the air binds to aromatic molecules released from flowers and the soil, and this helps the molecules bind more easily to scent-detecting cells in your nose. This accounts for the proverb, "Flowers smell best just before a rain."

F*lowers smell best just before a rain.*

Croaking frogs are another harbinger of summer rain. Being cold-blooded aquatic animals, frogs are most active when it's warm and humid. Even the wooden doors and windows in your house respond to an approaching storm. Wood fibers absorb water vapor in the air, making doors and windows swell and stick more in humid weather.

When natural clues portend a changing atmosphere, experienced weather-watchers look into the wind for signs of an approaching storm. Many major weather changes come with the wind, often from distant parts of the globe. Winds blowing into Missouri from the Caribbean bring weather with a tropical feel, while winds coming down from Canada bring an arctic bite.

The wind itself is a forecasting tool. The stronger the wind, the more rapid the change in weather is likely to be. Faster winds mean the atmosphere is experiencing greater pressure differences between one area and another. That may also mean that the storm, when it comes, will be more severe.

Frequent changes of wind direction with erratic cloud movements indicate an unstable atmosphere. Cloud movement tells you what winds are blowing above you. If the clouds above you are not moving in the same direction as the wind at ground level, that indicates that wind speed and direction are changing with altitude. This is a sign of an approaching storm.

Extending Your Senses

Using some simple and inexpensive instruments, you can extend your senses and gather more accurate measurements of the atmosphere. "A Home Weather Station" (page 104) tells you what you'll need.

By combining your observations and measurements with your knowledge of the weather patterns in your area, you can become an expert at interpreting your local weather forecast. Mary Miller's mother, who lives on a mountainside in central California, has been recording and graphing daily rainfall since 1984. By comparing her measurements with the records for the nearby valleys, she knows that when the weathercaster calls for an inch of rain in the valley, she'll likely get two to three inches at her house.

By paying attention to your own microclimate, you, too, can customize an official weather forecast. Official temperatures are often measured at the local airport. Because the airport has broad expanses of asphalt and few trees, it can be one of the hottest places in town on a sunny day. When the weathercaster predicts a scorching day, you can adjust those temperatures down to a more comfortable level if you live on a shady street with lots of trees and grass.

Easy Forecasting

If you assume that tomorrow's weather will be just like today's, you'll be right more than half the time. This is called *persistence forecasting* and it works best when the weather is changing gradually. For a slightly more sophisticated persistence forecast, you might assume that the moving weather patterns occurring today will continue tomorrow, predicting that the storm over you today will continue to move eastward, hitting the next state to the east tomorrow.

Another forecasting method uses past climate data to predict future weather. For instance, it typically rains three days out of 30 during the month of June in Flagstaff, Arizona. Therefore, if you forecasted no rain every day in June, you'd be right about 90 per-

A Home Weather Station

With your own weather station, you can make weather observations from your home or business. Here are some of the instruments you might want to get:

• **Anemometer and wind vane:** A wind vane indicates wind direction, and an anemometer, a rotating device with cups to catch the wind, allows you to measure wind speed.

• **Barometer:** Barometers measure atmospheric pressure (also known as barometric pressure). Atmospheric pressure is commonly reported in the media as inches of mercury, which refers to the height of a column of mercury in a mercury-tube barometer. A falling barometer indicates that stormy weather is more likely, while a rising barometer indicates clear skies ahead.

• **Thermometer:** There are rapid changes in temperature and wind when a warm or cold front is passing through. In cold climates during the winter, you can predict heavy snow if the barometer falls and the temperature rises, indicating an incoming warm front.

• **Psychrometer:** A psychrometer measures the humidity, which tells you the potential for dew, frost, or precipitation. When the relative humidity is above 70 percent, for instance, condensation and precipitation are much more likely than when the humidity is below 40 percent.

• **Rain gauge and snow gauge:** These devices measure the amount of rain or snow that has fallen.

If you establish a home weather station, you can join a network of 10,000 volunteers who provide twice-daily updates to the National Weather Service (NWS). Trained amateur storm- and tornado-spotters can also participate in the Skywatch Program, alerting the NWS whenever they see a dangerous storm brewing. For information on joining these programs, contact your local NWS office or emergency management agency.

cent of the time. This kind of forecasting, based on averages from past weather, is called *climatological forecasting.*

Neither persistence forecasting nor climatological forecasting require any special knowledge of atmospheric behavior. The yearly forecasts in the *Old Farmer's Almanac* are based on climatological forecasting, combined with astrology, folklore, and other "top secret" techniques. But persistence and climatological forecasts are often quite wrong because they don't take into account the highly variable nature of weather from day to day and year to year. As Tom Murphree reminds his students, "Climate is what you expect, weather is what you get."

Still, these easy forecasting methods set a minimum standard by which professional forecasts can be judged.

Professional Weather Forecasting

Like the forecasts you can make from your backyard, professional forecasts depend on making observations, recognizing weather patterns and how they evolve in space and time, and understanding the relationships among fundamental weather factors. These factors include temperature, pressure, winds, and the movement of water in and out of the atmosphere.

One of the biggest differences between personal and professional forecasting is that the pros have access to many more observations from many more places around the globe. Weather balloons and satellites determine temperatures, pressures, and winds high above the earth's surface. An international network of thousands of surface weather observation stations, including fully automated stations that operate 24 hours a day, provides data on cloud conditions, visibility, rainfall, temperature, wind direction, humidity, and pressure. All of these observations allow meteorologists to look at weather around the world and estimate how these conditions will travel and change. This global view of the weather is essential in making forecasts, especially those that predict the weather three or more days in advance.

Rules of Thumb

You can easily observe many weather patterns that indicate the weather that's likely to arrive in the next day or so. Here are some rules of thumb that work well in the midlatitudes:

- Clear skies and light winds at night are likely to lead to cold temperatures and dew or frost. But if the sky is cloudy and strong winds are blowing, the air will cool much less overnight.
- If cirrus clouds appear, followed by cirrostratus and then altostratus, a warm front is approaching and you may get some precipitation.
- If the wind shifts from the northwest to the west and then to the southwest, the temperature is likely to rise and the sky to become cloudier.
- If the wind shifts from the northeast to the north and then to the northwest, the temperature is likely to fall and the sky to clear.
- If tall cumulus clouds have formed by 11:00 A.M., showers or thunderstorms are likely by 3:00 P.M.

To predict the weather, you need to know what's heading your way. If you live in Cleveland, you could get a handle on approaching weather by phoning a friend in Detroit and asking her to describe the current weather conditions there. Most major weather systems travel roughly west to east across the United States, so chances are you'll experience Detroit's conditions within the next six hours. By making this phone call, you would be creating a very simple weather forecasting network.

In 1849, Joseph Henry, who aided Samuel Morse in the invention of the telegraph and became the first director of the Smithsonian Institution, revolutionized meteorology by establishing the nation's first communications network for reporting and forecasting the weather. In a bold use of new technology, Henry began distributing free weather instruments to telegraph offices, and extracted a promise from the operators to transmit current weather conditions every morning. He consolidated the information and started producing the first national weather maps. It wasn't long before a local newspaper, the *Washington Evening Star,* began using Henry's maps to print the country's first newspaper weather forecast.

By 1870, the U.S. War Department had taken over Joseph Henry's operation, attesting to the importance of weather forecasting for the military. (The U.S. Navy and Air Force still have their own weather observation and forecasting operations, separate from the government-funded weather service.) In 1891, the Department of Agriculture took over forecasting responsibilities and established the U.S. Weather Bureau. Eventually, the National Oceanic and Atmospheric Agency (NOAA) took over the Weather Bureau, which was rechristened the National Weather Service (NWS) in 1973.

Each day, millions of individual observations from across the U.S. and around the world flow into the NWS's National Meteorological Center in Maryland. The data are crunched by supercomputers to produce regional, national, and global forecasts. The Center distributes these forecasts to NWS field offices, which use the national forecast, along with data from their local areas, to come up with forecasts customized for their regions. The NWS forecasts reach most people through local print and broadcast media and Internet Web sites, such as www.nws.noaa.gov.

Radar Technology

The advent of modern weather forecasting depended not only on communication between distant locations, but also on the development and adaptation of three key technologies: radar, satellites, and high-speed computers.

Developed during World War II to detect and track aircraft, radar can also detect and track storms. Radar works by sending out radio waves, which reflect off obstacles in the air. The reflected waves provide information on what's out there and where it is, revealing, for example, where the heaviest rain is falling. Radar is particularly important for detecting thunderstorms and tornadoes that are too small or too far away to be observed by regular weather stations. The first generation of weather radar, installed in 1957, did an admirable job of detecting storms, rain, and snow, and served as the backbone of the nation's weather radar system for nearly 40 years.

In the 1990s, the National Weather Service installed a grid of new Doppler radars that covers most of the country. By detecting changes in the frequency of the reflected radio waves, Doppler radar can determine not only where a storm is, but also how it is moving. Doppler radar can detect the winds of an incipient thunderstorm or the beginnings of a tornado 20 minutes or more before it touches down, significantly increasing the warning time for people in its path. (Before Doppler radar was used, warnings were issued only after tornadoes were spotted on the ground.)

These satellite images show a midlatitude cyclone over the North Atlantic and Europe on March 30, 1994. In the infrared image (left), the coldest regions are white. Outer space is white as are the highest clouds. This makes some clouds more apparent in the infrared image than they are in the visible image (right).

Rainbow at Night

Over the centuries, people have come up with sayings, rhymes, and proverbs that summarize rules of thumb for predicting the weather.

You may have heard the rhyme: "Rainbow in the morning, sailor's warning. Rainbow at night, sailor's delight." You see a rainbow when the sun is at your back, shining on distant raindrops. If you see a rainbow in the morning, the storm producing the raindrops is to the west of you. In the midlatitudes, winds and storms generally travel from west to east. So a morning rainbow indicates a storm to your west, probably upwind and coming toward you. A rainbow in the late afternoon or evening is to the east of you, and the storm is probably moving away.

Eyes in the Sky

Weather satellites revolutionized the way meteorologists gather data. The first weather satellite, launched in 1960, proved its value in September 1961 when it provided dramatic images of powerful Hurricane Carla. These pictures helped convince 350,000 people along the Gulf Coast to seek shelter inland, the largest evacuation up to that time.

Satellites provide broad coverage of the oceans and remote land areas that affect our weather, much of which was previously unmonitored. On the West Coast, satellites over the North Pacific reveal approaching storms several days before they arrive. Since tropical cyclones usually form in remote areas of the ocean, satellites play a big role in tracking them.

Satellite images, a staple of the evening TV weather broadcast, come in two basic varieties: visible and infrared. Visible images are like photographs; both use visible light to create a picture. Infrared images create a picture using the infrared radiation that's continuously emitted by the ocean, land, air, and clouds—radiation that can't be detected by human eyes.

Dew on the Grass

When dew is on the grass, rain will never come to pass. Next time you're out early in the morning, look for drops of dew sparkling on the grass. If there's dew, chances are good that the weather will be fair.

Dew forms when the grass is cold enough that water in the air condenses on the blades of grass. When the surface of the earth (including the grass) cools, dew forms on the cool surface. It may seem strange, but dew is more likely to form when the air is relatively dry. That's because water vapor in the air helps keep the earth's surface warm at night. The earth's surface cools more on nights when the sky is clear, the total amount of water vapor in the air is low, and the air is calm. These conditions often accompany high pressure, fair-weather systems.

Warm things put out more infrared radiation than cold ones. So an infrared image is basically a picture of the earth's temperature. On an infrared image, the warmer areas are darker and cooler areas are lighter. Clouds that are low in the troposphere, where it's relatively warm, are darker than higher clouds. The lightest, brightest areas in the images are where towering cumulonimbus clouds reach the top of the troposphere and where heavy rain and snow are likely. Colorized infrared images are what you usually see on a TV weather report. Unlike visible images, they show you the weather during the night, as well as by day.

Both visible and infrared weather images reveal a lot about what's going on and what's coming up. For instance, a midwestern cold front with heavy snow and rain might appear as a comma-shaped swirl of bright white clouds stretching from Arkansas to Michigan. Afternoon thunderstorm clouds over the Rockies show up as a ragged line of puffy clouds that last for several hours and then fade away. A hurricane looks like a big, white cinnamon roll, often with a little hole, the eye of the storm, in the middle.

The Sun in His House

T he Zuni of New Mexico say: "When the sun is in his house, it will rain soon." What the Zuni refer to as a house, we call a halo, a ring of light that surrounds the sun or moon. Halos form when sunlight refracts off ice crystals in cirrus clouds. The high cirrus clouds responsible for the halo may be a sign that a warm front is moving in, bringing low clouds and rain or snow.

Weather satellites also monitor land and sea surface temperatures and temperatures high in the atmosphere. On-board sensors can monitor the ozone layer, measure the concentration of greenhouse gases, and monitor the winds at many levels. These satellites provide a huge number of observations that would be difficult or impossible to get any other way.

Computer Power

Radar, satellites, and other instruments gather data about temperature, pressure, winds, humidity, and other factors collected from many different points in the atmosphere and on the earth's surface. Powerful supercomputers allow meteorologists to feed this data into ever-more complex and detailed simulation models that predict how atmospheric conditions will change over time. The computer models must take into account how all these factors interact with each other. In the models, these complex interactions are represented by a set of mathematical equations based on the laws of physics.

Supercomputers can zing along at more than 2 billion calculations per second. That may seem fast, but running a forecasting model requires all that power and more. To predict next week's weather, a model begins with the present weather conditions and calculates what they will be several minutes from now. Then the model takes the results of those calculations and calculates the weather several more minutes into the future. This process continues until the model has calculated the weather for the next few days or weeks.

Tricky Predictions

Better computers aren't the only thing needed to produce a more accurate forecast. Both the actual atmosphere and the computer models of the atmosphere involve many complex interactions in which small disturbances can lead to big reactions. A weak swirling of winds over the tropical Pacific can lead to a massive typhoon over China a week later. A slight warming of the waters in the Gulf of Mexico can help trigger thunderstorms and tornadoes over the Midwest. So it's very important to supply a forecast model with detailed and accurate data about the state of the atmosphere at the beginning of the computer run.

Small changes can easily push the atmosphere into an unpredictable, chaotic state. To estimate how unpredictable the atmosphere is, meteorologists may run the same computer model several times, using slightly different starting conditions. If the resulting forecasts look similar, then meteorologists can feel more confident about them.

If different runs of the model give very different results, meteorologists attempt to minimize errors by merging the results from many runs to come up with what's called an *ensemble forecast*. They may also try to determine in advance where the problems with initial conditions will be most serious so that they can make additional observations in those areas.

Edward Lorenz, a meteorologist at the Massachusetts Institute of Technology, coined the term "butterfly effect" to describe the atmosphere's sensitivity to small changes over time. As he whimsically put it, a butterfly flapping its wings in Beijing today can alter storm systems in New York next week. Although he probably never meant for this metaphor to be taken literally, the butterfly effect suggests that tiny, unavoidable errors in measurement and in the atmospheric models can lead to bigger and bigger errors as the forecast is extended farther into the future. A forecast predicting tomorrow's weather might be dead wrong, but forecasts of the weather for tomorrow or the next day are more likely to be accurate than forecasts of next week's weather. It's very difficult to produce detailed and accurate forecasts beyond about two weeks into the future. At that point, you reach what scientists call a *predictability barrier.*

The atmosphere is more predictable when the weather is evolving slowly and when familiar patterns are apparent. In summer and winter, when extreme weather settles in, the weather is more predictable. In spring and fall, when the atmosphere is making large adjustments from one extreme to the other, the weather is less predictable. As any resident of Boston can report, April can bring warm and sunny weather one day followed by freezing rain and ice the next.

Predicting where and how much precipitation will fall is especially tricky. Consider this relatively common scenario for a winter storm approaching the West Coast. The weather satellite picks up a cold front with a swath of thick, moisture-laden clouds moving eastward over the Pacific. As it progresses toward land, the system looks like it's on a direct path to San Francisco. The night before the clouds are due to pass over, the TV weathercaster predicts heavy rainfall in and around San Francisco. The next morning clouds cover the sky, but there's barely enough drizzle to require windshield wipers, and everybody blames the weatherman for blowing it yet again.

In reality, he didn't do such a bad job. The front did cross over San Francisco, producing clouds and some rain. But though frontal clouds can cover a large area, they produce rain on a smaller scale. Next time you see a TV weather report, take a look at the color radar images. The splotchy pattern of dark red shows where the heaviest rain is falling, and blue or light green shows where light rain is falling. It takes some special conditions to produce that heavy precipitation, including enough energy to lift a lot of moist air against gravity. This lifting is likely to occur in the mountains surrounding San Francisco, but it's much less likely over the lower

> I enjoy being a television forecaster. It's one of the few jobs that you can be absolutely dead wrong on a somewhat regular basis and remain employed.
>
> —Steve Pool,
> TV meteorologist

elevations of the city itself. A small change in the jet stream can also quickly steer the storm farther north or south, leaving San Francisco at the edge of the front, with less rain than predicted.

And here's another complication: The satellite photos show the storm moving in a straight path due east. But within the storm, winds are swirling from southwest to northeast. Within this swirling eddy, rainfall is focused in a narrow band along the edge of the storm. If you've ever watched the water in a rocky stream, you know that the eddies erratically spin up and spin down and change course. The same is true of storms, which are eddies in the atmosphere. This makes predicting where rain will fall particularly difficult.

Long-Range Forecasting

Predicting tomorrow's weather is tough, but predicting the weather six months from now is even tougher. Every fall, Tom Murphree, who studies why we get long periods of unusual weather, gets calls from reporters and farmers asking for predictions of the winter weather.

All over the country, late summer and fall are busy times for meteorologists who study unusual weather patterns and work on long-range forecasting—that is, forecasting the weather and climate a month or more into the future. Several months before winter arrives, many people want to know how to prepare for the season. Managers at utilities want to know how much oil and gas they'll need to meet their customers' demands. Construction companies need to know how winter weather may slow down or speed up their projects. Water resource managers across the western U.S. need to prepare for flood control and water storage for spring and summer irrigation. Farmers in Kansas want to know how the snow cover will affect their winter wheat crops.

Long-range forecasts are not as specific as shorter-term forecasts. For example, a long-range forecast might say that the likelihood of above-normal temperatures in much of the Midwest is high in the coming winter. The forecast doesn't say how high the temperature will be or on what days the temperature will be high.

That's impossible to predict. The many processes that create weather and climate are too interactive and chaotic to allow more detailed forecasts of events that are still many weeks or months in the future.

To produce a long-range forecast, meteorologists use information about the parts of the weather system that tend to be persistent and have major effects on weather. Some of the most important of these phenomena are surface conditions, especially snow cover, soil moisture, and ocean temperatures. For example, when the upper Midwest has unusually heavy winter snowfall, spring and early summer there are likely to be cooler than normal. That's because the snow cover is likely to last longer, reflecting more incoming sunlight and keeping the ground wet longer as the snow melts. That would keep the soil cooler than normal well into spring and early summer, which would help keep air temperatures cooler, too.

El Niño Events

Meteorologists pay particularly close attention to the tropical Pacific when preparing long-range forecasts. There, unusual sea surface temperatures may develop and persist for many months. These temperature anomalies may drastically alter the normal weather patterns around the world.

Take, for example, the winter of 1992–93. In the U.S., the Southwest had exceptionally high rain and snowfalls, and major flooding. At the same time, the Pacific Northwest was suffering through a drought.

The chain of events that led to this atypical weather could be traced to sea surface temperatures in the tropical Pacific. That winter, the sea surface was cooler than normal in the western part of the tropical Pacific and warmer in the central and eastern parts. Since tropical storms tend to form over the warmest water, these changes in the sea surface temperature led to fewer tropical storms than usual in the western Pacific and more storms than usual in the central and eastern Pacific. This unusual pattern of sea surface temperatures and

Groundhog Day

I have a very positive attitude toward life, but there is one day each year on which it's a little harder to get out of bed in the morning—February 2. The reason: It's Groundhog Day, and I know that *Good Morning America* and nearly every other national and local news program will have cameras waiting to record whether or not a groundhog sees its shadow. Inevitably, dozens of people will ask me, 'Spencer, how come a groundhog knows how long winter will last and you don't?' What they're really saying is, 'Spencer, how come you're dumber than a groundhog?'. . .

The truth is, groundhogs aren't smarter than meteorologists (though they may be cuter). Scientific studies have shown absolutely no correlation between a groundhog seeing its shadow and the length of the winter.

—Spencer Christian in *Can It Really Rain Frogs?*

storms is part of what's called an *El Niño event,* a closely linked set of unusual conditions in the tropical Pacific and atmosphere that can lead to dramatic weather changes all over the world.

Fishermen along the tropical Pacific coast of South America call this ocean warming *El Niño de Navidad,* or "the Christ Child" since it happens around Christmastime. Scientists shorten the name to El Niño and recognize that this coastal warming is just one small part of a grand rearrangement of the entire tropical Pacific and atmosphere that occurs about every two to seven years. In this rearrangement, the normal patterns of sea surface temperatures and storms change, and remain changed for about 12 months.

El Niño events affect locations thousands of miles from the tropical Pacific as changes in the tropical storms alter the winds blowing into and out of the tropics. The changes in the winds lead to disruptions in the normal path of the subtropical and midlatitude jet streams and storm tracks.

During the winter of 1992–93, an unusually persistent jet and storm track developed. It extended from near Hawaii into southern California, Arizona, and New Mexico, and brought an exceptional number of warm, wet storms out of the subtropics. Meanwhile, the jet and storm track that would normally have spent much of the winter bringing rain and snow to Oregon and Washington shifted north into northern British Columbia, leaving the Northwest dry.

La Niña Events

Three years later, in 1995–96, Oregon and Washington had extremely heavy rain and snow and devastating floods, while the Southwest went through one of its worst droughts in years. Once again, the extreme weather could be traced back to unusual ocean temperatures in the tropical Pacific, over 5,000 miles away.

In 1995–96, the tropical Pacific sea surface temperature and storm anomaly patterns were the reverse of what they are in an El Niño year: The western region was warmer and stormier than normal, while the central and eastern areas were cooler and less stormy. These reversed conditions are part of what's called a *La Niña event*. (*La Niña* is Spanish for "the girl" and scientists use the term to describe how this event is the opposite of an El Niño event.) Like El Niño events, La Niña events also occur about every two to seven years and last about 12 months.

The changes that a La Niña event brings to the tropical Pacific, and its remote impacts, can be just as large as those during an El Niño event. In 1995–96, the shifts in the jets and storm tracks brought many warm, moist winter storms and record amounts of rain and snow to Oregon and Washington, and unusually few and weak storms to the Southwest.

One reason that changes in tropical Pacific sea surface temperature have such a large impact on weather is that the tropical Pacific takes up a very big chunk of earth's surface, extending from Indonesia in the west to South America in the east, and from about 15°S latitude to 15°N latitude—an area of about 20 million square miles, or more than five times the area of the lower 48 U.S. states. This huge tropical region is a giant solar collecting panel, absorbing much of the energy that drives the world's weather.

The trade winds that blow westward and toward the equator over the tropical Pacific cause water warmed at the surface to also flow westward. This creates what's called the *western tropical Pacific warm pool,* an accumulation of especially warm (80–85°F) water near New Guinea and Indonesia. The warm pool is actually a layer that floats on the cooler water below. Near New Guinea, the layer is about 700 feet thick. This warm, thick layer makes the tropical western Pacific a tremendous reservoir of energy that fuels some exceptionally dramatic weather, including about 45 percent of the world's hurricanes. The intense storm activity helps create the East Asian jet stream, the strongest jet in the world.

El Niño and La Niña events have their biggest impacts on the U.S. from November to March.

As the trade winds create currents that pile up water in the warm pool, they produce a sea surface temperature difference, with the eastern waters being about 65–70°F, or 15°F cooler, than those in the west. The cooler water inhibits storm formation, so the eastern tropical Pacific is a relatively quiet region for tropical storms, compared to areas farther west. This makes the jets that flow north and south of the tropical eastern Pacific relatively weak, too.

What does all this have to do with El Niño and La Niña events? As long as the trade winds blow normally, these sea surface temperature, storm, and jet patterns hold steady. But if the trades change, then many other things can shift, too.

The trades are driven by the difference between the high pressure in the eastern Pacific and the low pressure near Indonesia. (See chapter 3.) At the beginning of an El Niño event, the high pressure in the east (near Tahiti, say) gets lower, while the low pressure in the west (near Indonesia) gets higher. As the pressure difference gets smaller, the trades weaken. This causes the tropical western Pacific to cool down, while the central and eastern Pacific warm up.

The chain of events that results in El Niño is roughly reversed for a La Niña, which starts with an increase in the east-west pressure difference and the trades.

Scientists haven't yet figured out the chain of events that precedes the pressure and trade wind changes. One possibility is a change in Indian Ocean temperatures. Another is a change in the snowfall over Tibet. A large Tibetan snowfall might cause Tibet to stay cool well into the spring and summer, slowing down the development of the Asian summer monsoons and causing weaker trade winds to blow across the Pacific and into Asia.

Each El Niño event is different in the tropical Pacific and, even more so, in its impacts outside the tropics.

These jet stream and storm changes that El Niño events and La Niña events trigger outside the tropics are greatest during the winter, when the temperature differences between the warm tropics and the cold higher latitudes are greatest. That's part of why El Niño and La Niña have their biggest impacts on the U.S. from November to March.

In Zimbabwe, Indonesia, Australia, Peru, Brazil, Mexico, the United States, and many other countries, the disruption of normal rain and snowfall brought about by these events can cause disastrous droughts or floods. The far-reaching impacts of El Niño and La Niña have led to international efforts to better understand what causes them, and how to predict their effects on weather. The main goal of these efforts is to produce long-range forecasts that will allow countries to prepare well in advance for the coming weather changes.

Good forecasts depend on thorough observations of the key regions of the ocean, atmosphere, and land that create El Niño and La Niña events and their impacts. In the tropical Pacific, an array of shore stations, balloons, planes, buoys, ships, and satellites keep watch on weather conditions. But there are still large areas, such as in and over the ocean between Hawaii and Mexico, where little data are being collected.

Even with much better observations, El Niño and La Niña forecasts would still have a lot of weaknesses. What people want most are forecasts that will give them one to several months to prepare for the regional effects of El Niño and La Niña events. But forecasts with that sort of lead time have a very high predictability barrier to contend with. (See page 113.)

Besides, many forecasts of El Niño and La Niña and their impacts rely on observations of what's gone on during past events. But each El Niño event is different in the tropical Pacific and, even more so, in its impacts outside the tropics. So predictions based on the past can be very iffy. This is especially true for places like central California and much of the southern Great Plains, where the impacts from one El Niño to the next can be quite different. In Monterey and San Francisco, where Tom and Mary live, past El Niño events have helped create some very wet, some very dry, and some normal winters.

The connection between El Niño and U.S. weather is an example of a *teleconnection,* a link between widely separated parts of the weather system, such as sea surface temperatures near Samoa and rainfall in Albuquerque, New Mexico. Teleconnections produce many of the persistent changes in weather and climate that long-range forecasts try to predict.

The Meteorologist's Best Friend

Despite all the advances in forecasting technology and understanding, interpreting weather still depends in part on personal observations and experience. Before his evening broadcast, TV weather reporter Steve Raleigh checks the thermometer on the station's rooftop and glances out the window. "The window is a meteorologist's best friend," he says. There are times when the National Weather Service predicts rain, but Steve can see that the skies are clearing to the west, so he adjusts his forecast for the next day. You can do the same thing by paying attention to the atmosphere outside your door and to professional weather forecasts, and by developing your own intuition about the weather patterns in your hometown.

A Modern Meteorologist

Television weathercasters are the public face of meteorology. Every day they stick their necks out and predict the future—and catch grief when they get it wrong.

A couple of hours before his first broadcast of the day, Steve Raleigh, evening meteorologist for KRON-TV in San Francisco, sits down with his producer and reviews current weather conditions. He has hundreds of different information sources to choose from, including National Weather

Service forecasts, satellite and radar images, 3-D weather maps, Internet Web sites, wire services, and specialized forecasts from private companies. He strings images and information together into a visual "script" using a computer program developed at the station. Right before he goes on the air, he enters current temperatures onto the weather map, adjusts his tie, and strides into the broadcast booth.

For his broadcast, Steve stands in front of a giant blue screen called a chroma key, which overlays his image with the computer graphics. For the next three minutes, Steve uses the cued images to describe what's currently going on and what's coming up in the forecast. Occasionally, he'll include a short segment describing the science behind the weather. His personal credo is to make weather easy to understand. "Most people don't want lessons from a professor," says Steve. "They want information to help them plan the next day or week."

Climate

Natural cycles and human influences

A S AN ATMOSPHERIC SCIENTIST, Tom Murphree is often approached at cocktail parties by people who want to discuss their personal observations about climate change. They often begin with, "It seems to me that . . . " and complete the sentence with a version of one of the following: The summers are warmer; the winters are colder; storms are more severe than they used to be; floods or droughts are more common than they were 10 years ago.

At this point in the conversation, Tom usually politely answers that he's not familiar with the weather fluctuations in the person's hometown, but he could check the figures. But by that time, the person is launching into a pet theory about why the climate is changing. The explanations range from New Age spiritual ("the Earth Goddess is angry with us"), to media-influenced ("global warming is wreaking havoc with the weather"), to vaguely scientific ("it's part of a cycle in nature").

The problem with all of these explanations is that the original suppositions are often based on faulty or incomplete memories (last summer may have been unusually warm, but three summers ago it was cooler than normal), or on a lack of understanding of the difference between normal fluctuations in weather and long-term climate change. Tom refrains from commenting on the temper of the Earth Goddess, but he often points out that last winter's weather is not scientific proof that the climate is changing.

Weather is the state of the atmosphere over the short term. Climate is the average weather over the long term.

The folks who talk to Tom at cocktail parties aren't alone in equating anomalies in weather with climate change. During the summer of 1988, temperature records in the U.S. were broken, and drought shriveled crops in the grain belt. Following the drought, the media was awash with global warming stories. When the U.S. experienced some severe winter storms in 1996 and 1997, newspaper and magazine writers were asking climate experts whether extreme weather events were evidence of a global climate change. But it takes more than a few isolated events to make a trend.

Weather is the state of the atmosphere over the short term. Climate is the average weather over the long term. When scientists discuss long-term climate changes, they're talking about trends that persist for many years—in some cases, several decades; in other cases, thousands of years. The climatological norm is an average based on a number of decades of observation. From year to year, or within periods that can last a decade or two, the average annual temperature, precipitation, and storm frequency can vary quite a bit, both locally and globally, from their norms. In other

Changing Temperature

T hese graphs show how the earth's past temperature has varied from the planet's present average overall surface temperature of about 60°F. Over the past 800,000 years, earth's

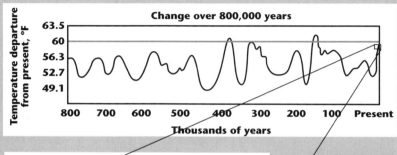

Change over 800,000 years

Temperature departure from present, °F

63.5
60
56.3
52.7
49.1

800   700   600   500   400   300   200   100   Present

Thousands of years

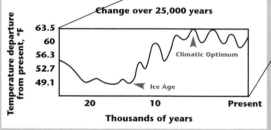

Change over 25,000 years

Temperature departure from present, °F

63.5
60
56.3
52.7
49.1

Climatic Optimum

Ice Age

20   10   Present

Thousands of years

temperature has shown glacial and interglacial periods, with major ice ages occurring roughly every 100,000 years over much of this time. A closer look at the past 25,000 years shows a warm period, the Climatic Optimum, about 7,000 years ago, and a cold period, the most recent ice age, about 18,000 years ago.

words, it's normal to have weather that's hotter or colder or wetter or drier than the climatological norm. It's also normal for the climate itself to change.

As we've been discussing throughout this book, weather is created by many complex, interacting processes. The interactions among these processes are always changing, sometimes creating rapid fluctuations of extreme weather and sometimes resulting in persistent weather patterns that can last for months or years. It can take decades to distinguish these natural variations from long-term trends or shifts in the climate.

Climate change is nothing new—it's been happening ever since the earth was formed. Our planet's climate has been slowly evolving, and most of the major changes have been gradual—too gradual for a person to observe in a single lifetime, or even in 10 lifetimes. The motion of the continents, the growth and decline of giant ice sheets, and the development of life itself, including the human species, have all affected the world's climate. The planet has gone through monumental changes, from barren rock to steamy marshlands teaming with dinosaurs, to vast expanses of glaciers. The last one billion years have seen alternating cooling and warming trends as the planet went through ice ages and temperate periods.

Scientists probe the geologic and atmospheric past to help them understand the conditions that affect the present and future climate. One hundred million years ago, for example, the atmosphere had higher concentrations of greenhouse gases (see page 19) and the world was a much warmer place than it is now. Those same gases are again increasing in the atmosphere, largely the result of human pollution, and many climate experts link that increase with a global rise in surface temperatures.

To investigate possible causes of climate change, including global warming, atmospheric scientists use complex computer models of the atmosphere. The models help determine the conditions that created past climates, and scientists use that understanding to make educated guesses about the future.

Ancient Climates

To figure out what the earth's climate has been like over the last billion years, geologists and climatologists have tallied the abundance and distribution of plant and animal fossils, used radioactive decay rates to date ancient lavas, soils, and ocean sediments, and looked for marks left by glaciers. From this and other evidence, they've discovered that the climate has gone through fairly large temperature swings. At one of its toastiest times, about 100 million years ago, the earth was as much as 28°F warmer than today. Those steamy times were the heyday of the dinosaurs.

In one of the earth's cool periods, around 450,000 years ago, the earth's higher latitudes were perhaps 15–20°F cooler than they are at present. The planet was in a major ice age, with much of the land surface of the Northern Hemisphere covered by glaciers. Since then, the earth's climate has fluctuated between periods of major ice expansion and contraction.

Since modern humans (*Homo sapiens sapiens*) appeared on earth, an estimated 45,000 to 90,000 years ago, they've experienced at least two major ice ages. The last one peaked about 18,000 years ago. Much of the earth's water was locked up in ice, and sea level may have been as much as 300 feet lower than it is today. The sea level drop uncovered a land bridge between Siberia and Alaska. Humans took advantage of that temporary route to migrate from Asia into North America.

The world at that time was a vastly different place than it is now: Ice covered most of present-day Canada, the northern U.S., and northern Europe. Much of the North Atlantic was frozen over. In the U.S., glaciers scraped across much of the Midwest, the Northeast, and down the slopes of the Rockies and the Sierra Nevada ranges, gouging out basins that would eventually become the Great Lakes and Yosemite Valley. Extensive spruce forests covered the Great Plains, and lush woodlands dotted with large lakes extended into what are now the western deserts. Large mammals, such as woolly mammoths and giant sloths, roamed near the ice's edge, supporting an expanding population of human hunters in Europe and North America. Even the tropics were affected, though the glaciers were far to the north. Fossil pollen records show that the Amazon basin was cool and dry with vegetation more typical of a savanna than today's lush rain forest habitat.

About 10,000 years ago, the planet had climbed out of the ice age and began experiencing an interglacial period, punctuated by periods of relatively mild warming and cooling. As the giant sheets of ice melted, sea levels rose rapidly, isolating the British Isles from the rest of Europe and flooding low-lying coastal regions around the world. Global surface temperatures continued to rise, peaking about 7,000 years ago in what's known as the *Climatic Optimum*, when summertime temperatures over much of the earth were several degrees Fahrenheit warmer than they are now. Since then, global temperatures have dipped and climbed several times.

Climate and Civilization

Throughout history, the fortunes of humans have been shaped by climate change. Centuries of warmer weather during the Climatic Optimum gave a boost to the development of human civilization. Heavier rains in the Middle East and North Africa helped make the development of widespread agriculture and livestock grazing possible. The first great agricultural civilizations of Egypt, Mesopotamia, and the Indus Valley flourished under this climate.

About 3,000 years ago, the climate in those areas became cooler and drier. The Egyptian and Babylonian civilizations survived by adapting to the changing climate (primarily by developing irrigation and food storage and distribution technologies), but other cultures, such as the Mycenaeans in Greece and the Hittites in Turkey, collapsed from drought and famine.

Patterns of rainfall in North America 7,000 years ago were also quite different than today. The heart of what is now the corn belt was much drier, with extensive sand dunes and sparse plant and animal life in present-day Nebraska. That constitutes a climate that's dramatically different from today's, yet the average summer temperature during the Climatic Optimum was only about 4°F warmer. This seemingly small temperature change was accompanied by an entirely different set of wind and rainfall patterns, creating a desert

Digging up the Past

Scientists interested in the earth's changing climate find some of their most useful clues by digging—into the ground, into ice, and into the sea floor.

Fossilized plant pollens are one clue to climate changes. By studying fossil pollen, scientists can learn where and when certain types of plants were living. For example, fossil pollen shows that several thousand years ago, spruce trees grew in North America far south of where they grow today, indicating a cooler climate.

The fossilized dens of pack rats living thousands of years ago are another source of climate information. True to their name, pack rats accumulate heaps of refuse (called *middens*). The fossilized vegetation found in these junk piles are indicators of the local climate at the time.

Ice samples taken from places like Antarctica and Greenland tell the history of the earth's climate through trapped air bubbles and dust. By chemically analyzing and dating the evidence, scientists can infer the past composition of the atmosphere and how the climate has changed. For instance, they have found that there was significantly less carbon dioxide in the air during periods when temperatures were cooler and ice was accumulating rapidly.

Sediment taken from the bottom of the ocean can also reveal climatic changes. The ratio of certain oxygen isotopes in plankton fossils can give clues about past ocean temperatures and whether ice caps were growing or shrinking when the plankton were living. This method of analysis has helped scientists identify climate shifts that took place hundreds of thousands of years ago.

in what is now the fertile high plains. The climate cooled again, and about 3,000 years ago, the Midwest more or less settled into its now-familiar pattern of humid summer weather and rains.

During the worldwide cooldown after the Climatic Optimum, there were also warm periods. During one of the warmest periods, from A.D. 800 to 1300, grapes thrived in England and oats in Iceland, and Canadian forests were north of their current ranges. With less ice in the North Atlantic, Vikings sailed to North America and also set up colonies in Greenland and Labrador. For a few hundred years, Norse settlers raised livestock and crops in these northern outposts.

Their fortunes took a difficult turn when the warm period ended and year-round ice and snow returned. From about 1600 to 1850, during what's called The Little Ice Age, worldwide weather was about 3°F cooler than now. The northern glaciers advanced and Europe's canals, rivers, and lakes regularly froze in the winter.

Climate changes have also dramatically affected people in the Americas. The Anasazi people in the American Southwest created extensive cliff dwellings about 1,000 years ago. During the warm period of the last millennium, when English vintners were fermenting wine from local grapes, the Anasazi prospered by growing maize. But in about A.D. 1280, the Anasazi suddenly abandoned their dwellings. By analyzing tree rings, scientists deduced that an extensive drought lasting a decade or so destroyed the maize crops and probably forced the Anasazi to leave their homes.

Causes of Change

Untangling the threads of climate change from the normal fluctuations in weather can be a tricky business. The Anasazi drought could have been caused by any number of factors, including small atmospheric changes and their chaotic repercussions. Changes in weather patterns can persist for a few months or even years. These patterns are often linked to changes in another part of the world. Unusually warm surface waters in the tropical eastern Pacific can affect rainfall patterns in Australia, South America, and the U.S. for several seasons or longer. (See pages 115–120.) Short-term cooling trends, lasting a year or more, often follow large volcanic eruptions. (See pages 22–23.) Such changes are not necessarily indicators of long-term global climate change.

Climatologists have also looked at variations in sunspots and other solar activity for clues about earth's warming and cooling trends. About every 11 years, there tends to be a peak in the number of sunspots and in the sun's output of ultraviolet radiation. The ebb and flow of the sun's activities, some scientists say, may contribute to global cooling and warming. Though The Little Ice Age corresponded with a period of very few sunspots, scientists have not found a convincing physical process by which solar changes could produce this or other major swings in earth's temperatures. Satellite measurements of the sun in the last two decades show that there has been less than half a percent change in solar radiation over the sunspot cycle, a change that's apparently too small to account for major climate changes.

Ice Ages

Still, climatologists agree that variations in solar radiation reaching earth have probably helped cause many climate shifts, particularly the ice ages. A widely accepted theory developed by Serbian geophysicist Milutin Milankovitch in the early 1900s links cyclic variations in the earth's motion to climate change. Milankovitch proposed that periodic fluctuations in earth's orbit and tilt, occurring in cycles of about 23,000, 41,000, and 100,000 years apart, could affect the amount of sunlight falling on the earth in different seasons and at different latitudes. The variations in earth's orbit affect how far the earth is from the sun, and the variations in the earth's tilt affect how sunlight is distributed over the surface. The timing of these cycles roughly matches the waxing and waning of ice advances.

Studies have confirmed that the amount of solar radiation falling in the Northern Hemisphere has varied seasonally by up to 9 percent over the last 100,000 years. Decreased radiation in the summer and increased radiation in the winter can encourage the growth of glaciers.

The Little Ice Age

As a child, you may have read the story called *Hans Brinker*. Written by Mary Mapes Dodge in 1865, it follows the adventures of a Dutch boy—Hans—living near Amsterdam in the mid-1800s. Hans and his friends spend most of the winter skating the canals near their home. In the end, the kind-hearted Hans gives up his chance to win a pair of silver ice skates in order to help a friend.

But if you were to visit the canals mentioned in the story, you might wait all winter and never see them freeze. Fat lot of good ice skates would be, silver or otherwise.

The frozen canals in *Hans Brinker* aren't just artistic license. It really was a lot colder back then. From the early sixteenth century through the mid-ninteenth century, there was a cool period known as The Little Ice Age. At that time, European winters were colder and harsher than they are today—cold enough for Holland's canals to regularly freeze over.

In the Northern Hemisphere, the scenario might go something like this: Suppose one winter, warmer ocean temperatures lead to more evaporation. Storm after storm dumps this extra moisture as snow on the northern continents. The large expanses of snow linger the following summer. Much of the solar energy that does reach the snowy surface is reflected back into space, keeping temperatures even cooler than they would otherwise be. The next winter, more snow accumulates, compressing the bottom layers of snow into ice.

For thousands of years, snow piles up, creating glaciers that inch toward the equator. The glaciers stop growing when the ocean cools and evaporation and snowfall decrease. The glaciers start to recede when the earth's orbit and tilt change again, increasing solar radiation in the summer. Summer sunshine melts the ice, exposing dark ground. The dark surfaces absorb more solar radiation and the melting process accelerates, forcing the glaciers to retreat back to the Pole. With less snow covering the ground, the earth warms up.

Fire or Ice?

Over the last million years, the earth's average temperature has gone up and down, ranging from about 4°F warmer than it is now to about 15°F cooler. Examination of past cycles of warming and cooling has led climatologists to think that the earth is just about ending an interglacial period and beginning another ice age. That means average global temperatures should be slowly dropping (although, as we pointed out earlier, individual years can still buck the trend and rise above average).

Average summertime temperatures had been dropping over the last 5,000 years or so. But that isn't happening anymore.

Since about 1850, the average global surface air temperature has risen and fallen, but overall it increased by about 1°F. The period from 1991 to 1995 saw the warmest global temperatures in recorded history. Forests and grasslands in the Northern Hemisphere are greening a week earlier than they did in the 1970s. Some places have warmed more than others: Western Antarctica's ice fields have warmed by 5°F in the last 50 years, possibly contributing to melting of sea ice and a worldwide rise in sea level.

As the planet has been heating up, the concentration of carbon dioxide in the atmosphere has been steadily increasing. People have been adding carbon dioxide and other so-called greenhouse gases to the atmosphere by burning carbon-based fuels, such as coal, oil, and gas. Is the increase in carbon dioxide forcing a global climate change? That question is being hotly debated. There's not universal agreement that carbon dioxide increases are solely to blame, but most climate experts would say that chances are good that human activities are connected to this recent global warming.

Global Warming and the Greenhouse Effect

As we discussed in chapter 1, solar energy is absorbed by the earth's surface and radiated upward as infrared radiation. As this infrared radiation travels through the atmosphere, some of it is absorbed and radiated back to the earth by the greenhouse gases, which keep the earth warmer than it would otherwise be. (See page 19.) The role of these gases in regulating earth's climate is called the greenhouse effect. For most of earth's history, the greenhouse effect has been in operation.

The most abundant and important greenhouse gases are water vapor and carbon dioxide, followed by methane, nitrous oxide, and chlorofluorocarbons (CFCs). Because of these gases, only about 5 percent of the outgoing infrared radiation can escape directly to space.

Carbon dioxide currently makes up only about 0.035 percent of the air. That translates to about 750 billion tons of atmospheric carbon. One hundred years ago, the atmosphere had only about 600 billion tons. But human activities, primarily burning fossil fuels in cars, factories, and power plants, have been increasing the level of carbon dioxide and other greenhouse gases in the atmosphere. In a century, carbon dioxide has increased by 25 percent. If current emission levels continue, the level of carbon dioxide will have doubled by some time in the next century.

Although the concentration of all the greenhouse gases in the atmosphere remains small compared with that of oxygen and nitrogen, many scientists fear that their rising levels may have significant effects on future climate. According to the Intergovernmental Panel on Climate Change (IPCC), an international group that included over a hundred scientists and economists, the projected doubling of atmospheric carbon dioxide will probably lead to further global warming. In their 1996 summary report, the IPCC anticipated that global temperatures would increase by anywhere from 1.8°F to 6.3°F by the year 2100. That could make the world warmer than it has been in at least the last 100,000 years.

Warming up by two degrees or even by six degrees may not seem like much. But the complex interactions between the sun, the atmosphere, and the earth's surfaces make it very difficult to predict exactly what the effects of this warming will be. Increasing the surface temperature of the earth could potentially affect nearly every aspect of the weather. Warming could disrupt current wind and rainfall patterns, storm frequency and intensity, ocean currents, ice cover, and sea level. These changes would affect natural habitats, agriculture, energy use, and human comfort and health.

Experimenting with a Model Planet

Predicting the climate consequences of warming is a daunting task, but atmospheric scientists are using a combination of observations, intuition, theory, and modeling to make their best estimates. The computer models used to predict changes in climate are not radically different from the models used to predict weather. But instead of predicting weather for the next few days, long-term climate models are used to help clarify the processes that cause climate changes and estimate likely scenarios many years into the future.

One of the biggest challenges scientists face when building climate models is deciding what to include in them. Stephen Schneider, a climatologist at Stanford University, was one of a large team of scientists who helped develop the climate model at the National Center for Atmospheric Research in Boulder, Colorado.

One of Schneider's interests has been assessing how life affects—and is affected by—climate change. Some of the biological factors that Schneider says need to be taken into account in climate models are ones you wouldn't normally think of. For instance, different plants evaporate water through their leaves at different rates, a process called *transpiration.* Broadleaf tropical plants transpire at faster rates than pine trees. This helps make the local climate in a rain forest more humid than that of a pine forest. Increased water vapor in the air affects rainfall and also increases the local greenhouse effect.

Biological effects, such as leaf transpiration rates, are included in the starting conditions for the models. Other inputs to the models include soil moisture, ocean currents, deep ocean temperatures, changing sea ice, glacier and snow cover, clouds, dust from air pollution, changes in the earth's orbit and tilt, and ocean salinity. In addition, the models include some of the same inputs as weather models, such as patterns of temperatures, pressures, humidity, precipitation, and winds. These and other starting conditions are plugged into the model, which calculates how they would evolve over time.

Making a climate model takes a team of researchers, including computer programmers and oceanic and atmospheric experts. Running a climate model often requires a super-fast computer and

Ozone

Ozone makes up less than .00005 percent by volume of the atmosphere, but it's critical for life on earth. Concentrated in a layer of the stratosphere 10 to 20 miles above the surface, ozone blocks much of the sun's dangerous ultraviolet radiation from reaching the ground. UV light damages DNA and cells, and causes sunburn, skin cancer, and cataracts.

By the mid-1970s, scientists suspected that ozone in the stratosphere was declining due to a class of atmospheric pollutants called chlorofluorocarbons (CFCs), used in refrigerants, industrial cleaners, and spray-can propellants. They assumed, however, that the loss was slow. Then, in 1985, the British Antarctic Survey reported a 40 percent decrease in springtime ozone concentrations over the South Pole. Ozone over North America and Europe has also decreased, although not as dramatically as it had over Antarctica and the southernmost portions of Argentina and Chile. The "ozone hole," actually a thinning of the stratospheric ozone layer, took the scientific community by surprise.

Ozone (O_3) is formed when a molecule of oxygen (O_2) combines with a free oxygen atom. Ozone production happens naturally in the stratosphere, driven by intense solar radiation. When CFCs make their way into the stratosphere, they break down into highly reactive by-products. These volatile chlorine compounds can steal an oxygen atom from ozone molecules, destroying the molecules' ability to absorb UV radiation.

Recognizing the need to deal with the thinning of the ozone layer, representatives from 56 nations signed the Montreal Protocol in 1987, a treaty to curb emissions of CFCs. Strengthened in 1990, the international agreement has begun to show some positive effects: Concentrations of ozone-harming compounds in the lower atmosphere are leveling off, even decreasing by some measurements. But it may take a decade or more before chlorine levels in the stratosphere start to decline and even longer before the ozone hole over Antarctica disappears.

weeks or months of computing time for each experiment. A typical experiment involves running the model twice. Using current conditions as a starting point, the climate model might calculate the climate 150 years into the future with levels of atmospheric carbon dioxide held constant. Then projections for the increased accumulation of greenhouse gases would be plugged in and the model would be run again. Scientists would compare the results from the two runs to tease out the possible consequences of a human-enhanced greenhouse effect.

It often takes a year or more to run a complete set of experiments. The experimental results include projections for temperatures, pressures, humidity, and winds. From this, cloudiness, solar radiation reaching the earth's surface, and infrared radiation are calculated. Specialists analyze the results to determine the projected changes in regional weather and climate and the impacts that those changes might have on such things as habitats, water resources, agriculture, and forest fires. They also assess the economic impacts on such things as energy use and insurance claims resulting from natural disasters.

It takes a broad community of workers to sift through the results of a climate model, Schneider says. Assessing the impacts on forest management, for instance, requires detailed knowledge of fire probabilities arising from various climate conditions. The foresters may find that the frequency of forest fires in some forests doesn't increase significantly with an average surface warming of two degrees. But the probability of fires may increase dramatically if afternoon temperatures rise above 85°F and the relative humidity drops below 55 percent for three weeks in a row.

Putting Models to the Test

Computer models can't duplicate the complexity of the atmosphere, and climatologists like Schneider are the first to admit that the climate models aren't perfect. "The results are a mixed bag," he says. "We're good at predicting events on the scale of seasons and ice ages. Our ability to pinpoint local or short-term events is terrible."

To test their models, a process called *validation,* scientists often go back and study past climates. If they plug the conditions that existed prior to the time of the dinosaurs into their models and run them out, the models should predict the balmy climate that existed when the giant lizards ruled the earth. The point is not to use the models to determine what past climates were like—that's done with fossils and other geologic evidence. The point is to test the model by plugging in the past conditions, running them out, and seeing if the model can reproduce the past climate. If that works, you know your model is good, Schneider says.

Science, my boy, is made up of mistakes. But they are mistakes that are useful to make because they lead little by little to the truth.

—Professor Otto Liedenbrock, fictional character in the Jules Verne novel, Voyage to the Centre of the Earth

Volcanoes have been surprisingly useful for validating climate models. "Volcanoes are nature's gift to the climatology world," Schneider says. "They allow you to calibrate your instruments. When you adjust your model to take away a specific amount of energy—say, two watts per square meter—from the sun [by shading the earth with clouds of volcanic dust], your model better cool by the amount that we observe in nature." After the eruption of Mount Pinatubo, Jim Hansen of NASA's Goddard Institute for Space Studies predicted that global temperatures would decrease by a few tenths of a degree. Later observations bore him out.

As scientists learn more about how the atmosphere works, they adjust their models. With each new piece of information, they increase the accuracy and utility of the models.

Climate modelers learn from their mistakes. For instance, some early models didn't account for the cooling effects of aerosols, microscopic bits of dust and other particles that partially shade the sun, potentially slowing greenhouse warming. These cooling effects should be greatest in the Northern Hemisphere, where aerosol-emitting industries are concentrated. Recent models have taken

The amount of carbon dioxide (CO_2) in the atmosphere has been increasing for many years. Using observations taken near the top of the Mauna Loa volcano in Hawaii, this graph shows the increase over the last 40 years. During this time, there was a 15 percent increase in CO_2. (The CO_2 concentration units are parts per million by volume.) The level of carbon dioxide in the atmosphere naturally varies within each year, dropping in the Northern Hemisphere's spring and summer, when plant photosynthesis and carbon dioxide uptake are high, and increasing in the Northern Hemisphere's fall and winter, when plant photosynthesis is low and more carbon dioxide is released through plant decay.

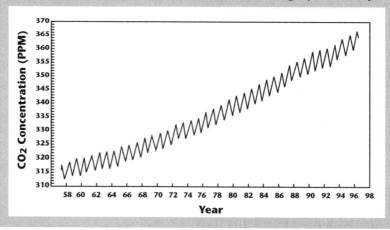

this regional aerosol cooling into account, predicting less warming in the Northern Hemisphere. Temperature records from the 1970s and '80s show some of the regional and seasonal patterns predicted by the models, an indication that the models are not way off base.

Challenges and Uncertainties

Clouds are one of the greatest sources of uncertainty in climate models. Higher temperatures can translate into more evaporation and hence more clouds.

Clouds influence the climate in two major ways: They can cool the surface by reflecting solar radiation from above, and they can warm the surface by slowing down the escape of infrared radiation from below. Which effect dominates depends in part on the altitudes of the clouds. Low clouds block incoming sunlight more than they reduce outgoing infrared radiation, so they have a net cooling effect. High clouds can have the opposite effect, tending to warm more than they cool. Clouds' thickness also has an effect on whether they warm or cool the surface. The models must therefore take into account and predict the height and thickness of clouds, as well as cloud coverage and other complexities of cloud physics.

The interactions between the ocean and the atmosphere also pose major difficulties for climate modelers. In early models, the ocean was treated as a giant mass of water that changed slowly. But the real ocean is constantly responding to shifting winds, surface warming, and changing currents. Climatologists are building complex models that link the atmosphere and the ocean and simulate how they influence each other. But these models still have a long way to go. For example, they have a difficult time reproducing the trade wind and sea surface temperature changes seen during El Niño and La Niña events, two major natural climate variations that involve many complicated interactions between the ocean and the atmosphere. (See pages 115–120.)

Searching for Climate Fingerprints

As scientists analyze observations from the field and pore over the results of their model runs, they look for what are called *climate fingerprints*. These fingerprints are patterns in the physical evidence that link the theories to actual climate changes that have happened in the present or past.

Scientists are searching for fingerprints by looking for changes that match those predicted by the models. For example, the models indicate that increased greenhouse gases will not just cause an increase in the earth's average surface temperatures; they will cause a set of linked changes in stratosphere temperatures, nighttime versus daytime temperatures, and polar versus tropical temperatures. Some climate models and theories predict that increasing carbon dioxide will cause surface temperatures to go up, but stratospheric temperatures to go down. Cooling of the stratosphere has been observed, but by more than scientists expected, indicating that other factors may be contributing to the cooling effect. Climatic fingerprints could indicate that a pattern observed in nature—surface warming, stratospheric cooling, more frequent or severe storms and droughts—fits into a model-predicted pattern of persistent and global climate change.

Not everyone accepts the results offered by climate models. "Some people have trouble believing them," says Schneider, "because the factors that cause climate change in the past are not identical to what will happen in the future. For example, the current rate of atmospheric modifications is very fast. We don't have past metaphors for that rate of change."

Biological Impacts

Suppose everyone in the scientific, political, and public sphere accepted the climate model indications that the climate will continue to warm, from 2–6°F by the end of the next century. "What's the big deal?," you might ask. After all, on a typical summer day you may experience 20 or more degrees of warming from morning to afternoon, with no apparent ill effects. In fact, most living things can tolerate high short-term variability, as long as temperatures don't cause freezing or heat exhaustion. The problem comes when the temperature change is sustained over time. As Stephen Schneider says, "You or I may enjoy a sauna for a half hour, but try it overnight and you'd be very dead."

Most biological systems are fine-tuned to their environment and climate and may be sensitive to changes in average temperature of only a few degrees. During the last ice age, when the earth was about 10°F cooler than at present, the varieties of trees that are now found in Canada were growing in South Carolina. Those tree species, along with all the species of birds, mammals, insects, and other living things that depend on them, had thousands of years of gradual climate change to migrate to South Carolina with the cooling temperatures, and thousands of years more to move back up to Canada after the ice retreated. If current projections of global warming occur, the change will be so swift that biological systems may not have enough time to respond without significant disruption. "The trees of South Carolina can't call the van lines and just move the entire ecosystem," Schneider says.

During past transitions from ice age to warmer interglacial periods, different species migrated north at different rates. The differing rates of migration disrupt the groupings of plants and animals that make up an ecological community, and probably contributed to past extinction of some species.

What About the Weather?

Climatologists are also concerned about how global warming may change global and regional weather patterns, and the impacts those changes may have on people and the environment. Global warming could cause large changes in ocean currents, winds, and precipitation, leading to far-reaching and sustained effects on the environment in general. Here are just a few scenarios that have emerged from climate model studies. These are theoretical impacts, not fully validated predictors about future weather.

• **Rainfall patterns:** With warming, increased evaporation and precipitation are likely. Since 1910, precipitation in the U.S. has increased by about 10 percent. But the change has not been uniform. Several climate models suggest that in some areas, global warming will lead to an increase in precipitation of 15 percent in winter and a decrease of 5 to 10 percent in summer. Currently, the places where rain naturally falls—in river basins, for instance—are seeing more intense episodes of rain, more "gully washers." Many models predict more of these locally intense rains, which could spell trouble for levees, dams, and reservoir systems that have been overburdened by flooding in recent years.

• **Sea level:** In many models, the Poles warm significantly more than the tropics. Warming of polar ice could begin to melt the edges of the ice caps and raise sea level. Water also expands when heated, raising sea levels further. But warming could also cause more evaporation and more snowfall, which might increase the snowpack in the middle of the polar ice caps, partially offsetting the melting at the edges. Which process will dominate is under debate. Some models project that water expansion and ice melting could result in a sea level rise by the year 2100 of a few inches up to several feet.

• **Agriculture:** Increasing levels of carbon dioxide in the atmosphere could be beneficial for crops; warming could also open many higher-latitude areas to agriculture. Crops now grown in Nebraska might be shifted to southern Saskatchewan, for instance. But warmer, drier conditions could increase the water needs of Nebraska farmers and might sharply reduce their crop yields. Farther south in Mexico and other developing nations, the impact could be even more severe, crippling some farmers and ranchers who lack the resources and infrastructure to adapt readily to change.

• **Ocean circulation:** Changes in precipitation and runoff in the upper latitudes could change the salinity of the North Atlantic and alter the world's ocean currents. Salty water is denser and sinks below fresher water. The sinking of cold, salty water in the far North Atlantic helps drive much of the ocean's circulation. Ocean currents, such as the Gulf Stream, have major impacts on climate. Some climate models indicate that small changes in the salt balance of the North Atlantic could trigger large changes in ocean currents, sea ice, and climate.

• **Extreme events:** With global warming, more heat waves and fewer cold snaps may be likely. With increased evaporation from warming, hurricanes and winter snowstorms might also be more intense and more frequent. With hotter temperatures comes a higher chance of wildfires.

• **Desert expansion:** Global warming might intensify the Hadley circulation, leading to more intense high pressure and less precipitation in desert areas. Added to the desert expansion already observed from human activities, such as overgrazing by livestock and deforestation, this warming-enhanced desertification could expand existing deserts or create new ones.

Climate Surprises

In addition to the scenarios described above, climatologists have posited some rapid and radical climate changes called *climate surprises,* such as global warming of 10°F or more over a few hundred years. Many of these extreme events involve hard-to-predict changes in ocean circulation.

Stephen Schneider cites an example of a climate surprise that happened about 12,000 years ago when the Northern Hemisphere was recovering from its recent glacial advance. Warm-weather plants were just getting reestablished in Europe, when, within a few hundred years, there was a dramatic return to ice-age conditions. This shift was characterized by the reappearance in Europe of tundra flowers with the Latin name *Dryas octopetella,* which is why this mini ice age is called The Younger Dryas. For about a thousand years, much of northwestern Canada and most of Europe went through a mini ice age before they warmed up again.

How did this happen? Scientists think it may have started when large quantities of arctic ice melted and dumped a flood of fresh water into the North Atlantic. Fresh water floats and freezes more readily than salty water. The fresh water helped block the formation and sinking of cold, salty water, which drives the inflow of warm water from the south. During the winter, the fresh water may have frozen and formed a large expanse of sea ice that also blocked warm waters from the south. As a result, much of Europe and Canada went back into an ice age.

According to Stephen Schneider, another Younger Dryas event is very unlikely under any current global warming scenario. But it does raise the very real possibility that complex feedback systems, like those among ice caps, ocean currents, and the atmosphere, can lead to rapid climate changes. The more rapidly we alter the basic components of our climate, Schneider says, the greater the probability of more surprises in the future.

What to Do?

People who are skeptical that carbon dioxide pollution will cause global warming often point out the shortcomings of climate models and uncertainties about their results. They say that the atmosphere is too complex to accurately model many years into the future, or to separate natural variations from those caused by humans. They also point out that it is difficult to tell just when and where the effects will be felt. But many climate scientists think that we're already seeing the smudged fingerprints of a human-induced warming, pointing to the global warming of the last century, stratospheric cooling, and the increasing frequency of extreme weather events.

Critics argue that the models can't prove that global temperature trends or individual weather events are the result of human-enhanced greenhouse effect, and they caution against hasty judgments. Before possibly disrupting the economy by increasing taxes on fossil fuels or forcing a reduction in greenhouse gas emissions, they say, we should wait for proof that human activities are having a significant and detrimental effect on the climate.

That's a minority opinion among most climate experts. At the 1992 Earth Summit in Rio de Janeiro, developed nations agreed to cut their releases of greenhouse gases back to 1990 levels by the year 2000. The Europeans took an early lead by proposing firm target reductions and timetables. The United States, which produces about 20 percent of the world's total greenhouse gas pollution, more than any other nation, is taking a more conservative approach, proposing relatively modest reductions in emissions, funding new technologies, and calling for more energy conservation. Since the summit, however, the U.S. has increased its emission of carbon dioxide by 10 to 20 percent.

The truth is that just about everyone on the planet contributes in varying degrees to the buildup of greenhouse gases. Cars and trucks in the U.S., coal-burning factories in Russia and China, and wood fires in India all produce atmospheric carbon pollution. Taking action will involve money, will, and greater efforts to conserve energy, build low-pollution power plants, expand the use of public transportation, and increase the use of renewable energy sources.

Future Weather

Perhaps we can take encouragement for the future by looking back at the past. Throughout its history, the earth has experienced much larger fluctuations in climate than humans are likely to cause, yet it has remained a livable planet.

Climate observations, theories, and models will continue to improve, but we'll never predict the future with complete accuracy—the atmosphere is too chaotic. But the more we study the atmosphere, the better we'll get at anticipating the range of future possibilities. As long as we keep looking, we'll constantly be discovering surprising things about the environment. That's the nature of science. Sixty years ago, meteorologists knew very little about jet streams; today, monitoring ridges and troughs in these upper-atmosphere winds has become a critical part of weather forecasting. Perhaps in the next few years, scientists will uncover a feedback mechanism that tempers global warming and heads off widespread climate change. Whether or not that happens, Tom Murphree suspects that 20 years from now we'll look back on at least a dozen major discoveries about processes that affect weather, climate, and global change. These discoveries will answer questions that scientists are currently asking, and lead to questions that they don't yet know enough to ask.

Sources

Bohren, Craig F. *Clouds in a Glass of Beer, Simple Experiments in Atmospheric Physics.* New York, NY: John Wiley & Sons, Inc., 1987.

Bohren, Craig F. *What Light Through Yonder Window Breaks? More Experiments in Atmospheric Physics.* New York, NY: John Wiley & Sons, Inc., 1991.

Burroughs, William J., Bob Crowder, Ted Robertson, Eleanor Vallier-Talbot, and Richard Whitaker. *The Nature Company Guides: Weather.* Alexandria, VA: Time Life Books, 1996.

Christian, Spencer, and Antonia Felix. *Can It Really Rain Frogs? The World's Strangest Weather Events.* New York, NY: John Wiley & Sons, Inc., 1997.

Demillo, Rob. *How Weather Works.* Emeryville, CA: Ziff-Davis Press, 1994.

Denis, Jerry. *It's Raining Frogs and Fishes.* New York, NY: Harper Collins Publishers, Inc., 1992.

Dott, Robert H., Jr., and Donald R. Prothero. *Evolution of the Earth.* New York, NY: McGraw-Hill, Inc., 1994.

Dunlop, Storm, and Francis Wilson. *Weather and Forecasting.* New York, NY: Collier Books, 1982.

Feynman, Richard P. *Lectures on Physics.* Reading, MA: Addison-Wesley Publishing Company, 1963.

Graedel, T. E., and P. F. Crutzen. *Atmosphere, Climate and Change.* New York, NY: Scientific American Library, 1995.

Hewitt, Paul G. *General Physical Science.* New York, NY: Harper Collins, 1994.

Houghton, J., L. Meira Filho, B. Callander, N. Harris, A. Kattenberg, and K. Maskell. *Climate Change 1995: The Science of Climate Change.* New York, NY: Cambridge University Press, 1996.

Junger, Sebastian. *The Perfect Storm.* New York, NY: W. W. Norton & Co., Inc., 1997.

Keen, Richard A. *Skywatch East: A Weather Guide.* Golden, CO: Fulcrum, Inc., 1992.

Keen, Richard A. *Skywatch: The Western Weather Guide.* Golden, CO: Fulcrum, Inc., 1987.

Lehr, P. *Weather: Air Masses, Clouds, Rainfall, Storms, Weather Maps, Climate.* New York, NY: Simon and Schuster, 1987.

Lockhart, Gary. *The Weather Companion, An Album of Meteorological History, Science, Legend, and Folklore.* New York, NY: John Wiley & Sons, Inc., 1988.

Ludlum, D., R. Holle, and R. Keen. *Clouds and Storms.* New York, NY: Knopf, 1995.

Lutgens, Frederick K., and Edward J. Tarbuck. *The Atmosphere, An Introduction to Meteorology.* 6th ed. Englewood Cliffs, NJ: Prentice-Hall, 1995.

Lyons, Walter A. *The Handy Weather Answer Book.* Detroit, MI: Visible Ink Press, 1997.

Moran, Joseph M., Michael D. Morgan, and Patricia M. Pauley. *Meteorology, The Atmosphere and the Science of Weather.* 5th ed. Upper Saddle River, NJ: Prentice-Hall, Inc., 1997.

Morgan, Michael D., and Joseph M. Moran. *Weather and People.* Upper Saddle River, NJ: Prentice-Hall, Inc., 1997.

Schneider, Stephen H. *Global Warming.* San Francisco, CA: Sierra Club Books, 1989.

Schneider, Stephen H. *Laboratory Earth.* New York, NY: Harper Collins, 1997.

Schneider, Stephen H., and Randi Londer. *The Coevolution of Climate & Life.* San Francisco, CA: Sierra Club Books, 1984.

Williams, Jack. *The Weather Almanac.* New York, NY: Vintage Books, Random House, 1994.

Williams, Jack. *The Weather Book.* New York, NY: Vintage Books, Random House, 1992.

Credits and Acknowledgments

Design by Gary Crounse
Illustrations by Esther Kutnick and Randy Comer
Edited at the Exploratorium by Pat Murphy
Production editing by Ellyn Hament
Design production by Stacey Luce
Sidebars on pages 11, 31, 43, 58, 88, 90, 92, 95, 128, and 131 by Pearl Tesler

Image Credits

Pages 2, 26, 50, 58, 76, 100, and 122: illustrations by Randy Comer; pages 6, 16, 18, 29, 42, 45, 47, 52, 53, 66, 71, and 81: illustrations by Esther Kutnick; pages 84–85: from *The Atmosphere—An Introduction to Meteorology* by Frederick K. Lutgens and Edward J. Tarbuck, reprinted with permission; page 97: courtesy of the University of Wisconsin at Madison, Cooperative Institute for Meteorological Satellite Studies; page 108: courtesy EUMETSAT; page 121: courtesy KRON-TV; page 138: CO_2 concentration figure courtesy of Charles D. Keeling and Timothy Whorf, Scripps Institution of Oceanography, University of California at San Diego, La Jolla, California.

Acknowledgments from the Authors

Mary Miller thanks her husband, Jeff Schaffer, for his love and support; editor Pat Murphy for her patience, confidence, and problem-solving abilities; Stephen Schneider for long hours spent explaining the politics, science, and significance of global warming; Steve Raleigh for letting her hang around the TV station and for explaining how local weathercasts are created; Spencer Christian for his personal and professional views on national weathercasting; Jeff Cruzan for his insights on the special properties of water; her father, Hugo Miller, for inspiring an interest in science and answering questions about flying in bad weather; her mother, Mary Miller, and sister, Cynthia Fine, for providing weather anecdotes and moral support. She also thanks all the people, too numerous to list, who contributed vivid, real-life stories about tornadoes, hurricanes, wind storms, blizzards, ice storms, lightning, and squall-line thunderstorms.

Tom Murphree thanks his mother, Emilee Murphree, and father, Harold Murphree, for all their encouragement of his curiosity and for helping him, in the end, do things his way; Huug van den Dool of the Climate Prediction Center for sharing many wonderful hours walking the beach, bicycling, and deliberating on the ways nature works; Deidre Sullivan of Hartnell and Monterey Peninsula Colleges for sharing her fascination with the atmosphere; Sharon Roth Franks of Scripps Institution of Oceanography for being so darn smart and funny; Mark Boothe of the Naval Postgraduate School, Frank Schwing of the Pacific Fisheries Environmental Laboratory, and Don McManus of the American Meteorological Society for spending many hours pondering through thought experiments with him, and showing him even better ways to make science sensible; Pat Murphy of the Exploratorium for battling many fierce storms and making this whole project happen; and coauthor Mary Miller for helping him see so many new stories in science.

Acknowledgments from the Exploratorium

At the Exploratorium, no one works alone. This book would not exist without the efforts of many members of the Exploratorium staff and friends of the Exploratorium.

Thanks to Gary Crounse for his design work and pataphysical serenity in moments of panic; to Pearl Tesler for her heroic efforts in research and writing; to Esther Kutnick and Randy Comer for their fine illustrations; to Megan Bury for her photo research and assistance with all the squirrelly details; to Stacey Luce for her production work and smiling acceptance in times of crisis; to Ellen Klages for her meticulous copyediting; to Ellyn Hament for her valiant efforts to catch every last error; to Larry Antila for his courier services; to Ruth Brown for her (as always) sage advice, moral support, and editing acumen; to Kurt Feichtmeir for his cheerful support, budgeting acumen, and constant efforts to smooth the waters; to Rob Semper for his continuing support of our publishing enterprise; and to our scientific advisors, Exploratorium physicists Paul Doherty and Tom Humphrey, for their wide-ranging expertise, exciting weather stories, and thoughtful review of drafts. Special thanks to David Sobel at Henry Holt and Company for his editing expertise.

And of course, I'd like to thank our authors, Mary Miller and Tom Murphree, for their perseverance and dedication.

—Pat Murphy

Index

Air. *See* Atmosphere
Air masses, 78, 79
  movement of, 79–80
  semipermanent, 89–90
Air pressure. *See* Atmospheric
  pressure
Alberto, Tropical Storm, 96
Altitude, air temperatures
  and, 28. *See also* Elevation
Altocumulus clouds, 66, 67
Altostratus clouds, 67
Andrew, Hurricane, 95
Anemometer, 104
Anticyclones, 86–87
  semipermanent, 89–90
Atlantic Ocean
  Bermuda or Atlantic High
    over, 89
  climate change predictions,
    142
Atmosphere
  cloud formation in, 63–65
  components of, 17, 19
  compression warming, 28
  elevation and density of,
    21–22
  expansion cooling in, 28
  layers in, 18
  sublimation in, 62–63
  temperature and, 17–22
  turbulence in, 38
  water in, 53–54
  water transport in, 60–61
Atmospheric pressure, 29–30
  See also Pressure gradients
  elevation and, 21
  measuring, 104
  signs of change in, 101–102
  vertical winds and, 34
  wind predictions, 30–33

Barometer, 104
Beaufort scale, 37
Bergeron process, 72
Bermuda High, 89
Blizzards, 76–77, 82
Butterfly effect, 112

California Current, 13, 15, 56

Carbon dioxide
  climate change and, 128,
    132, 133
  Earth Summit agreement
    and, 144–145
  graph of variations, 138
  infrared radiation
    absorbed by, 19, 23
Carla, Hurricane, 109
Chinooks, 28, 36, 70
Chlorofluorocarbons (CFCs),
  133, 135
Cirrocumulus clouds, 66, 67
Cirrostratus clouds, 67
Cirrus clouds, 65, 66, 67, 80
Claudette, Tropical Storm, 96
Climate, 122–145
  civilization and, 127–129
  defined, 123
  elevation and, 22, 70
  global circulation cells
    and, 46
  historical variations, 124,
    125–132
  modeling changes in, 125,
    134, 136–140
  predictions, 141–143
  relative humidity and,
    54–56
  weather vs., 105, 123–125
Climate surprises, 141
Climatic Optimum, 124, 127,
  128, 129
Clouds
  categorization of, 65–69
  climate change modeling
    and, 138–139
  as forecasting tool, 102
  formation of, 63–65,
    69–71
  funnel, 94, 95
  low pressure areas and, 30
  satellite images of, 110
  seeding of, 73
  thunderstorm formation
    and, 91
  in troposphere, 18
  upslope winds and, 35, 36
  with warm front, 80

water cycle and, 51, 53
Cold fronts, 80–82
Compression warming, 28
Condensation
  temperature of air and,
    53–54, 61, 62
  water cycle and, 51, 53
Condensation nuclei, 63, 72,
  73, 80
Convection, described, 25
Cooling, 28, 55–56
Coriolis effects, 40–42
Cumulonimbus clouds, 66,
  68, 69
  hail formation in, 73
  lightning and, 88
Cumulus clouds, 64, 65, 66,
  68–69
Cyclones, 96
  See also Hurricanes;
  anticyclones and, 86–87
  midlatitude, 83, 84–86
  semipermanent, 89–90

Dew point, 58, 61, 62
"Doldrums," 43–44
Downbursts, 92
Downwinds and upwinds,
  34–36, 38
Drought
  blocking patterns and, 87
  El Niño/La Niña and, 119

Earth
  axial tilt, 5, 6, 7
  energy budget of, 16, 22–23
  infrared radiation and, 20
  orbital variations and
    climate of, 130, 131
  spin of, Coriolis effects and,
    40, 42
  temperature of, 16, 20,
    124, 132
Earth Summit agreement,
  144–145
East Asian jet stream, 118
El Niño, 115–117
  climate modeling and,
    139

cut-off highs/lows in, 89
local weather and, 3
midlatitude cyclones and,
83, 86–87
Pacific Ocean and, 118
rapids and whirlpools in,
89–90
weather forecasting and,
83, 85, 114
winter storms and, 83,
86–87, 89
zonal flow by, 89

Lake-effect snow, 75
La Niña, 89, 117–120, 139
Latent heat, 52
dew point and, 62
storm formation and, 78
in thunderstorms, 93
in tropical cyclones, 97
water cycle and, 51, 53
Lightning, 88
cloud types and, 68–69
Linda, Hurricane, 97
Little Ice Age, 129, 130, 131
Low pressure area, 29
wind predictions and, 30–33

Microclimates, 17, 24, 103
Modeling climate change,
125, 134, 136–140
predictions based on,
141–143
Monsoons, 39
Coriolis effect and, 40–41,
42
Mountains
clear air turbulence over,
38
cloud formation over,
69–70, 71
hailstorms and, 73
as moisture barrier, 57
precipitation on flatlands
vs., 69–70
pressure gradients in, 32,
33
rapid temperature change
in, 20–21
vertical wind patterns in,
35–36

National Weather Service
(NWS), 104, 107

Nimbostratus, 66, 67, 68
Nor'Easter storms, 79
Northern Hemisphere
Buys-Ballot's trick in, 47
Coriolis effects and winds
in, 42
polar jet stream, 49
seasons in, 6
trade winds in, 43

Oceans
air masses formed on, 79
air temperatures and, 11,
56
blocking pattern origins
in, 89
climate change studies,
128, 139
climate surprises and, 143
cloud formation and,
64–65
currents and temperature
of, 12–13, 15, 56
density of air over, 29–30
hurricane formation over,
96, 99
Old Farmer's Almanac, 105
Ozone, 18, 135
Ozone layer, satellite
monitoring of, 111

Pacific air masses, 79
winter storms and, 83
Pacific Ocean
air masses formed on, 79
El Niño, 115–117
La Niña, 117–120
typhoons in, 96, 99
Persistence forecasting, 103,
105
"Pineapple Express," 60
Polar air masses, 79
Polar cells, 46
Polar jet stream, 49
Precipitation
cloud location and, 67
cloud types and, 67, 68,
71
formation of, 72–73
global warming predic-
tions, 141, 142
midafternoon, 35, 36, 70, 74

on mountains vs. flat
lands, 69–70
"rain shadow" and, 70
seasonal variations in,
74–75
water cycle and, 51, 53
Predicting climate change,
141–143
Pressure gradients, 31–32
air mass movement and,
80
long-distance winds and,
39, 43, 48
midlatitude cyclones and,
83, 86
Psychrometer, 58, 59, 104

Radar, forecasting and,
107–108
Rain
See also Precipitation
formation of, 72–73
freezing (glaze), 73
global warming predic-
tions, 141
historical patterns,
127–129
in hurricanes, 96
long-distance winds
and, 39
in thunderstorms, 91
upslope winds and, 35–36
water cycle and, 51, 53
words for, 75
Rainbows, 109
Rain gauge, 104
"Rain shadow," lower
precipitation in, 70
Relative humidity, 54
elevation and, 59
heat stress index
and, 55
indoors, 60
measuring, 58–59
temperature and, 61–62
Ridge
blocking pattern by, 87
isolated (cut-off high), 89
in jet stream, 48, 86
Rocky Mountains
clear air turbulence, 38
upslope winds in, 35–36
winter storms in, 83

About the Authors

Mary K. Miller is a senior science writer at the Exploratorium, which feeds her abiding interest in the science of everyday life. She also writes about biology, earth sciences, chemistry, and psychology for magazines such as *Earth, The Sciences, Popular Science,* and *Reaction Times.*

Dr. Tom Murphree is a professor of meteorology at the Naval Postgraduate School in Monterey, California, where he conducts research on the physics of climate—especially the interactions of the ocean, atmosphere, and land that cause El Niño and La Niña events. Tom also directs Global Systems Integration, a business specializing in integrated science education for students and teachers.

About the Exploratorium

The Exploratorium, San Francisco's museum of science, art, and human perception, is a place where people of all ages make discoveries about the world around them. The museum has over 600 exhibits, and all of them run on curiosity. You don't just look at these exhibits—you experiment with them. At the Exploratorium's exhibits, you can play with a captive tornado, generate an electric current, see what's inside a cow's eye, and investigate hundreds of fascinating natural phenomena.

Each year, over half a million people visit the museum. Through programs for teachers, the Exploratorium also encourages students to learn by asking their own questions and experimenting to find the answers. Through publications like this one, the Exploratorium brings the excitement of learning by doing to people everywhere.

Visit the Exploratorium's home page on the World Wide Web at:

http://www.exploratorium.edu

Next time you are in San Francisco, come visit the Exploratorium!